T0251611

APPLIED SINGULAR
INTEGRAL EQUATIONS

APPLIED SINGULAR
INTEGRAL EQUATIONS

B N Mandal

NASI Senior Scientist
Indian Statistical Institute
Kolkata, India

A Chakrabarti

NASI Senior Scientist
Indian Institute of Science
Bangalore, India

  **CRC Press**
Taylor & Francis Group
an **informa** business
www.crcpress.com

6000 Broken Sound Parkway, NW
Suite 300, Boca Raton, FL 33487
270 Madison Avenue
New York, NY 10016
2 Park Square, Milton Park
Abingdon, Oxon OX 14 4RN, UK

 Science Publishers
Enfield, New Hampshire

Published by Science Publishers, P.O. Box 699, Enfield, NH 03748, USA
An imprint of Edenbridge Ltd., British Channel Islands

E-mail: _info@scipub.net_   Website: _www.scipub.net_

Marketed and distributed by:

CRC Press
Taylor & Francis Group
an **informa** business
www.crcpress.com

6000 Broken Sound Parkway, NW
Suite 300, Boca Raton, FL 33487
270 Madison Avenue
New York, NY 10016
2 Park Square, Milton Park
Abingdon, Oxon OX 14 4RN, UK

Copyright reserved © 2011

ISBN: 978-1-57808-710-5

CIP data will be provided on request.

The views expressed in this book are those of the author(s) and the publisher does not assume responsibility for the authenticity of the findings/conclusions drawn by the author(s). Also no responsibility is assumed by the publishers for any damage to the property or persons as a result of operation or use of this publication and/or the information contained herein.

All rights reserved. No part of this publication may be reproduced, stored in a retrieval system, or transmitted in any form or by any means, electronic, mechanical, photocopying or otherwise, without the prior permission of the publisher, in writing. The exception to this is when a reasonable part of the text is quoted for purpose of book review, abstracting etc.

This book is sold subject to the condition that it shall not, by way of trade or otherwise be lent, re-sold, hired out, or otherwise circulated without the publisher's prior consent in any form of binding or cover other than that in which it is published and without a similar condition including this condition being imposed on the subsequent purchaser.

Printed in the United States of America

Preface

Integral Equations occur, in a natural way, in the course of obtaining mathematical solutions to mixed boundary value problems of mathematical physics. Of the many possible approaches to the reduction of a given mixed boundary value problem to an integral equation, Green's function technique appears to be the most useful one and, such Green's functions involving elliptic operators (Laplace's equation being an example) in two variables, are known to possess logarithmic singularities. The existence of singularities in the Green's function associated with a given boundary value problem, thus, brings in singularities in the kernels of the resulting integral equations to be analyzed in order to obtain useful solutions of the boundary value problems under consideration.

The book covers a variety of linear singular integral equations, with special emphasis on their methods of solution. After describing the various forms of integral equations in the introductory chapter (chapter 1), we have broken up, the whole material presented in the book, into nine chapters. In chapter 2, simple elementary methods of solution of the famous and most important Abel integral equation and its generalizations have been discussed first, and, then the singular integral equations of the first kind which involve both logarithmic as well as Cauchy type singularities in their kernels have been taken up for their complete solutions. The theory of Riemann-Hilbert problems and their applications to solutions of singular integral equations involving Cauchy type kernels has been described in a rather simplified manner, in chapter 3, avoiding the detailed analysis, as described in the books of Gakhov and Muskhelishvilli (see the references at the end). Particular simple examples are examined in detail to explain the underlying major mathematical ideas. Some very special methods of solution of singular integral equations have been described in chapter 4, wherein simple problems of various types are examined in detail. The chapter 5 deals with a special type of singular integral equation, known as hypersingular integral equations, along with their occurrence and utility in solutions of mixed boundary value problems arising in the study of scattering of surface water waves by barriers and in fracture mechanics.

Hypersingular integral equations of both first as well as second kinds have been examined with special emphasis on problems of application to physical phenomena. Both analytical as well as approximate methods of solution of such integral equations have been described in this chapter. Some special approximate methods of solution of singular integro-differential equations have been explained in detail, in connection with simple problems, in chapter 6. This particular chapter, like a major portion of the material described in chapter 5, is the result of some recent research having been carried out by the authors and other workers. The chapter 7 deals with the Galerkin method and its application. In chapter 8, numerical methods of solution of singular integral equations of various types have been explained and some simple problems have been discussed whose numerical solutions are also obtained. The error analysis in the approximate as well as numerical methods of solution of singular integral equations studied in the chapters 5 and 6, of the book has been carried out to strengthen the analysis used. The final chapter 9 involves approximate analytical solution of a pair of coupled Carleman singular integral equations in semi-infinite range arising in problems of water wave scattering by surface strips in the form of inertial surface and in the form of elastic plate, which have been studied by the authors and coworkers recently.

It is hoped that the book will help in picking up the principal mathematical ideas to solve singular integral equations of various types that arise in problems of application. It is further hoped that even though all the ideas are explained in the light of specific simple problems of application, there is no lack of rigor in the analysis for readers and users looking for these aspects of singular integral equations. It should therefore serve as a book, which helps in introducing the subject of singular integral equations and their applications to researchers as well as graduate students of this fascinating and growing branch of applied mathematics.

B. N. Mandal
A. Chakrabarti

February 2011

Contents

Chapter 1

Introduction

In this introductory chapter we describe briefly basic definitions concerning *integral equations* in general, and *singular integral equations* in particular. Integral equations arise in a natural way in various branches of mathematics and mathematical physics. Many initial and boundary value problems associated with linear ordinary and partial differential equations can be cast into problems of solving integral equations. Here we present some basic definitions and concepts involving singular integral equations and their occurrences in problems of mathematical physics such as mechanics, elasticity and linearised theory of water waves

1.1 BASIC DEFINITIONS

An equation involving an unknown function $\varphi(x)$ with $a \leq x \leq b$ (a,b being real constants), is said to be an integral equation for $\varphi(x)$, if $\varphi(x)$ appears under the sign of integration. A few examples of integral equation is given below:

Example 1.1.1

$$\int_a^b K_1(x,t)\, \varphi_1(t)\, dt = f_1(x), \ a \leq x \leq b$$

where $K_1(x,t)$ and $f_1(x)$ are known functions and $\varphi_1(x)$ is the unknown function to be determined.

Example 1.1.2

$$\varphi_2(x) + \int_a^b K_2(x,t)\, \varphi_2(t)\, dt = f_2(x), \ a \leq x \leq b$$

where $K_2(x,t)$ and $f_2(x)$ are known functions and $\varphi_2(x)$ is unknown.

Example 1.1.3

$$\varphi_3(x) + \int_a^b K_3(x,t) \left[\varphi_3(t)\right]^2 dt = f_3(x), \ a \le x \le b$$

where $K_3(x,t)$ and $f_3(x)$ are known functions and $\varphi_3(x)$ is unknown.

The known functions $K_1(x,t), K_2(x,t), K_3(x,t)$, appearing in the above equations, are called the *kernels* of the integral equations involved, and the other known functions $f_1(x), f_2(x), f_3(x)$, are called the *forcing terms* of the corresponding integral equations. We emphasize that integral equations whose forcing terms are zero, are called *homogeneous* integral equations, whereas for *nonhomogeneous* integral equations, the forcing terms are non-zero. The function $K_i(x,t), f_i(x), \varphi_i(x)$ $(i = 1, 2, 3)$ appearing in the above examples are, in general, complex-valued functions of the real variable x.

The integral equations in the Examples 1.1.1 and 1.1.2 above are examples of *linear* integral equations, since the unknown functions φ_1, φ_2 there, appear *linearly*, whereas the integral equation in the Example 1.1.3, in which the unknown function appears *nonlinearly*, is an example of nonlinear integral equation. In the present book we will consider, only linear integral equations.

Some further examples of integral equations involving either functions of several real variables or several unknown functions are now given.

Example 1.1.4

$$\varphi_4(\mathbf{x}) + \int_\Omega K_4(\mathbf{x},\mathbf{t})\varphi_4(\mathbf{t}) \, dt = f_4(\mathbf{x}), \ \mathbf{x} \in \Omega \subset \mathbb{R}^n, \ n = 2, 3 \dots.$$

Here $\varphi_4(\mathbf{x})$ is the unknown function of several variables $x_1, x_2, \dots, x_n (n \ge 2)$ and the Kernel K_4 as well as the forcing term f_4 are known functions. This is an example of a linear integral equation in an n-dimensional space $(n \ge 2)$.

Example 1.1.5

$$\varphi_i(x) + \sum_{j=1}^N \int_a^b K_{ij}(x,t) \, \varphi_j(t) \, dt = f_i(x), \ a \le x \le b, \ i = 1, 2, \dots, N.$$

Here the set of functions $\varphi_i(x)$ $(i = 1, 2, \dots, N)$ is an unknown set, and the kernel functions $K_{ij}(x)$ as well as the forcing functions $f_i(x)$ are known.

This is an example of a *system* of N one-dimensional linear integral equations.

In the present book we will be concerned with only those classes of integral equations, which are known as *singular integral equations,* and for such equations, the kernel function $K(x,t)$ has some sort of *singularity* at $t = x$. A singularity of $K(x,t)$ at $t = x$, when it exists, may be *weak* or may be *strong*. For a weak singularity of $K(x,t)$ at $t = x$, the integral

$$\int_a^b K(x,t)\,\varphi(t)\,dt$$

for $a < x < b$ exists in the sense of Riemann while for a strong singularity of $K(x,t)$ at $t = x$, the integral

$$\int_a^b K(x,t)\,\varphi(t)\,dt$$

for $a < x < b$ has to be defined suitably.

Linear integral equations may be of *first* or *second* kind. A first kind integral equation has the form

$$\int_a^b K(x,t)\,\varphi(t)\,dt = f(x),\ a \le x \le b \tag{1.1.1}$$

while a second kind integral equation has the form

$$\varphi(x) + \lambda \int_a^b K(x,t)\,\varphi(t)\,dt = f(x),\ a \le x \le b \tag{1.1.2}$$

where λ is a constant.

If both limits of integration a and b in (1.1.1) and (1.1.2) are constants, then these equations are called integral equations of *Fredholm type*, whereas, if any one of a and b is a known function of x (or simply equal to x), then the corresponding integral equations are said to be of *Volterra type*.

If the kernel $K(x,t)$ is continuous in the region $[a,b] \times [a,b]$ and the double integral

$$\int_a^b \int_a^b |K(x,t)|^2\,dx\,dt$$

is finite, then the integral equations (1.1.1) and (1.1.2) are called regular integral equations.

Although the equation (1.1.2) is the standard representation of Fredholm integral equation of second kind, there exists another form of equation given by

$$\mu\varphi(x) + \int_a^b K(x,t)\, \varphi(t)\, dt = f(x),\ a \le x \le b \qquad (1.1.3)$$

where it is evident that $\mu = \dfrac{1}{\lambda}$ and is absorbed in the forcing term. One advantage of this representation is that, on setting $\mu = 0$, one gets the first kind Fredholm integral equation.

Example 1.1.6

$$\varphi(x) - \lambda \int_0^1 e^{x-t}\, \varphi(t)\, dt = f(x),\ 0 \le x \le 1,$$

where λ is a known constant. This is an example of a nonhomogeneous Fredholm integral equation of the second kind.

Example 1.1.7

$$\varphi(x) - \int_0^x xt\, \varphi(t)\, dt = f(x),\ 0 \le x \le 1.$$

This is an example of a nonhomogeneous Volterra integral equation of second kind.

As mentioned above, singular integral equations are those in which the kernel $K(x,t)$ is unbounded within the given range of integration. Based on the nature of unboundedness of the kernel, one can have *weakly singular* integral equation, *strongly singular* integral equation and *hypersingular* integral equation.

If $K(x,t)$ is of the form

$$K(x,t) = \frac{L(x,t)}{|x-t|^\alpha}$$

where $L(x,t)$ is bounded in $[a,b] \times [a,b]$ with $L(x,x) \ne 0$, and α is a constant such that $0 < \alpha < 1$, then the integral $\int_a^b K(x,t)dt\ (a < x < b)$ exists in the sense of Riemann, and the kernel *is weakly singular*, and the corresponding integral equation (first or second kind) is called a *weakly singular integral equation*. Also the logarithmically singular kernel

$$K(x,t) = L(x,t) \, ln \, |x - t|$$

where $L(x,t)$ is bounded with $L(x,x) \neq 0$, is also regarded as a weakly singular kernel.

Example 1.1.8

$$\int_0^x \frac{\varphi(t)}{(x-t)^{1/2}} \, dt = f(x), \, x > 0$$

where $f(0) = 0$. This is an example of a nonhomogeneous Volterra equation of first kind with weak singularity. This is in fact the Abel integral equation attributed to the famous mathematician Niels Henrik Abel (1802–1829) who obtained this equation while studying the motion of particle on smooth curve lying on a vertical plane.

Example 1.1.9

$$\int_a^b \varphi(t) \, ln \, \left| \frac{t+x}{t-x} \right| \, dt = f(x), \, a < x < b.$$

This is a first kind Fredholm integral equation with logarithmically singular kernel. This integral equation occurs in the linearised theory of water waves in connection with study of water wave scattering problems involving thin vertical barriers.

If the kernel $K(x,t)$ is of the form

$$K(x,t) = \frac{L(x,t)}{x-t}, \, a < x < b$$

where $L(x,t)$ is a differentiable function with $L(x,x) \neq 0$ (the function L can still be weaker!), then the kernel $K(x,t)$ has a *strong* singularity at $t = x$, or rather it has a *singularity of Cauchy type* at $t = x$, and the integral $\int_a^b K(x,t)dt$ is to be understood in the sense of *Cauchy principal value (CPV)*, as denoted and defined by

$$\rlap{\textstyle\int}{\textstyle-}_a^b K(x,t) \, \varphi(t) \, dt = \lim_{\varepsilon \to +0} \left[\int_a^{x-\varepsilon} K(x,t) \, \varphi(t) \, dt + \int_{x+\varepsilon}^b K(x,t) \, \varphi(t) \, dt \right]$$

where the cut indicates that the integral is defined as CPV only. The corresponding integral equation is called a *Cauchy-type singular* integral equation.

Example 1.1.10

$$\fint_a^b \frac{\varphi(t)}{t-x}\, dt = f(x),\ a < x < b$$

where the integral is in the sense of CPV. This is called a nonhomogeneous Cauchy type singular integral equation first kind. This integral equation occurs in various types of physical problems. It can be solved only when the behaviour of $\varphi(t)$, dictated by the physics of the problem in which it occurs, at the end points is known.

If the kernel $K(x,t)$ is of the form

$$K(x,t) = \frac{L(x,t)}{(t-x)^2}\ ,\ a < x < b$$

where $L(x,t)$ is continuous and $L(x,x) \neq 0$, then $K(x,t)$ has a *very strong singularity* at $t = x$.

For simplicity, we choose $L(x,t) \equiv 1$, then $K(x,t) = \dfrac{1}{(t-x)^2}$, and the integral $\displaystyle\int_a^b \frac{\varphi(t)}{(t-x)^2}\, dt \ (a < x < b)$ *cannot* be defined in Riemann sense.

However, it can be defined in the sense of Hadamard finite part of order 2, as denoted and defined by

$$\times \int_a^b \frac{\varphi(t)}{(t-x)^2}\, dt = \lim_{\varepsilon \to +0} \left[\int_a^{x-\varepsilon} \frac{\varphi(t)}{(t-x)^2}\, dt + \int_{x+\varepsilon}^b \frac{\varphi(t)}{(t-x)^2}\, dt - \frac{\varphi(x+\varepsilon) + \varphi(x-\varepsilon)}{\varepsilon} \right]$$

where the cross before the integral indicates that the integral is defined in the sense of Hadamard finite part.

Example 1.1.11

$$\times \int_{-1}^1 \frac{\varphi(t)}{(t-x)^2}\, dt = f(x)\ \ -1 < x < 1$$

where $\varphi(\pm 1) = 0$ and the integral is defined in the sense of Hadamard finite part. This is an example of a first kind hypersingular integral equation.

Note: The cut in the integral sign to denote a Cauchy principal value integral and the cross before the integral sign to denote a hypersingular integral will not be used further.

Remark: There exists in the literature a huge amount of work related to *second kind* integral equations of Fredholm type for regular kernels since the solution is unique. However there exists no such work for first kind integral equations of Fredholm type for *regular kernels* since the solution need not be unique. This is illustrated by considering the first kind integral equation

$$\int_0^1 (x+t)\,\varphi(t)\,dt = 1, \ \ 0 \le x \le 1$$

Here the kernel $K(x,t) = x+t$ is obviously a regular kernel. Solution of this integral equation is not unique. It is easy to see that it has solutions

$$\varphi(t) = -6 + 12t, \ \ \varphi(t) = -24t + 36t^2.$$

In fact any number of solutions can be obtained if one follows the method used in Chakrabarti and Martha (2009) for finding approximate solutions of Fredholm integral equations of the second kind. Thus there is no point to consider finding solutions of first kind Fredholm integral equations with regular kernels. However, this is not the case with first kind integral equations with *singular* kernels. A considerable part of this book is devoted to solving first kind singular integral equations analytically as well as approximately.

We now demonstrate the occurrence of singular integral equations (weakly singular, Cauchy singular and hypersingular) in varieties of problems of mathematical physics, like classical mechanics, elasticity and fluid mechanics.

1.2   OCCURRENCE OF SINGULAR INTEGRAL EQUATIONS

1.2.1   Weakly singular integral equation (Abel's problem)

We consider, first, the problem in classical mechanics, which is that of determining the time a particle takes to slide freely down a smooth fixed curve in a vertical xy-plane (see Figure 1.2.1), from any fixed point (X,Y) on the curve to its lowest point (the origin 0).

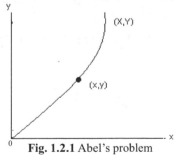

Fig. 1.2.1 Abel's problem

If m denotes the mass of the particle and $x = \psi(y)$ denotes the equation of the smooth curve where ψ is a differentiable function of y, then we obtain the energy conservation equation as given by

$$\frac{1}{2} mv^2 + mgy = mgY \tag{1.2.1}$$

where v is the speed of the particle at the position (x, y) at time t, assuming that the particle falls from rest at time $t = 0$ from the point (x, y), and g represents acceleration due to gravity.

We can express the relation (1.2.1) as

$$\frac{ds}{dt} = -\left[2g(Y - y)\right]^{1/2} \tag{1.2.2}$$

by using the arc-length $s(t)$, measured from the origin to the point (x, y), where a minus sign has been used in the square root since s decreases with time t during the fall of the particle.

Using the formula

$$\frac{ds}{dy} = \left[1 + \{\psi'(y)\}^2\right]^{1/2} \tag{1.2.3}$$

where $\psi'(y) = \dfrac{d\psi}{dy}$, we can express (1.2.2) in the form

$$\frac{dy}{dt} = \frac{dy}{ds}\frac{ds}{dt} = -\left[\frac{2g(Y - y)}{1 + \{\psi'(y)\}^2}\right]^{1/2} \tag{1.2.4}$$

and, this on integration, gives

$$\int_Y^0 \left[\frac{1 + \{\psi'(y)\}^2}{2g(Y - y)}\right]^{1/2} dy = -\int_0^T dt = -T \tag{1.2.5}$$

where T is the total time of fall of the particle, from the point (x, y) to the origin $(0,0)$.

Writing

$$\varphi(y) = \left\{\frac{1 + \{\psi'(y)\}^2}{2g}\right\}^{1/2} \tag{1.2.6}$$

the relation (1.2.5) can be written as

$$\int_0^Y \frac{\varphi(y)}{(Y-y)^{1/2}}\, dy = T = f(Y), \text{ say, } 0 < Y < a. \qquad (1.2.7)$$

Note that $f(0) = 0$. We thus find that the time of descent of the particle, T, can be determined completely by using the formula (1.2.7), if the shape of the curve $x = \psi(y)$, and hence the function $\varphi(y)$ is known.

If we consider, alternatively, the problem of determination of the *shape of the curve*, when the time of fall $T(= f(Y))$ is known, which is the historic Abel's problem, then the relation (1.2.7) is an integral equation for the unknown function $\varphi(y)$, which is known as *Abel's integral equation* or simply Abel integral equation.

The most general form of Abel integral equation is given by

$$\int_0^x \frac{\varphi(t)}{\{h(x)-h(t)\}^\alpha}\, dt = f(x), \ x > 0 \ \left(f(0)=0\right) \qquad (1.2.8)$$

where $h(x)$ is a monotonically increasing function and α is a real constant such that $0 < \alpha < 1$.

We note that the equations (1.2.7) and (1.2.8) are linear Volterra integral equations of the first kind. These are weakly singular integral equations.

In the equation (1.2.8) if we put

$$h(x) = x^2, \ h(x) = \sin x \ (0 < x < \pi/2), \ h(x) = 1 - \cos x \ (0 < x < \pi/2),$$

then we obtain some special Abel integral equations.

Abel discovered the equation (1.2.7) in 1826, and thus the year 1826 may be regarded as the year of birth of the topic integral equation.

1.2.2 Cauchy type singular integral equation

A. A crack problem in the theory of elasticity

The mathematical problem to determine the distribution of stresses, in two dimensions, around a Griffith crack, can be shown to be reducible to a singular integral equation involving a Cauchy-type singular kernel, as described below.

Using Cartesian xy co-ordinates, with $|x| < 1$, $y = 0$ representing the crack in an infinitely elastic plate, this basic mixed boundary value problem of the theory of plane elasticity can be posed as follows. To solve the equations of equilibrium, as given by

$$\left.\begin{array}{l} \dfrac{\partial \sigma_{xx}}{\partial x} + \dfrac{\partial \sigma_{xy}}{\partial y} = 0 \\[4mm] \dfrac{\partial \sigma_{xy}}{\partial x} + \dfrac{\partial \sigma_{yy}}{\partial y} = 0 \end{array}\right\} \quad \text{for } y > 0 \qquad (1.2.9)$$

under the conditions

$$\left.\begin{array}{l} \sigma_{xy} = 0 \text{ on } y = 0,\ -\infty < x < \infty \\[2mm] \sigma_{yy} = -p(x) \text{ on } y = 0,\ |x| < 1, \\[2mm] u_y = 0 \text{ on } y = 0,\ |x| > 1 \end{array}\right\} \qquad (1.2.10)$$

where u_x, u_y represent the displacement components, and $\sigma_{xx}, \sigma_{xy}, \sigma_{yy}$ represent the stress components at the point (x, y), it being assumed that there are no body forces.

We also force that the derivatives of u_x, u_y tend to zero as $(x^2 + y^2)^{1/2} \to \infty$.

Then, as shown by Sneddon (1994), we can utilize the following representations of the displacement and stress components at point (x, y):

$$\left.\begin{array}{l} u_x = \dfrac{1}{\sqrt{2\pi}\,\mu} \displaystyle\int_0^\infty \dfrac{P(\xi)}{\xi}\, (1\text{-}2\eta\text{-}\xi y)\, e^{-\xi y} \sin\xi x\, d\xi \\[5mm] u_y = \dfrac{1}{\sqrt{2\pi}\,\mu} \displaystyle\int_0^\infty \dfrac{P(\xi)}{\xi}\, \{2(1\text{-}\eta)+\xi y\}\, e^{-\xi y} \cos\xi x\, d\xi \end{array}\right\} \qquad (1.2.11)$$

and

$$\left.\begin{array}{l} \sigma_{xy} = -\sqrt{\dfrac{2}{\pi}}\, y \displaystyle\int_0^\infty \xi\, P(\xi)\, e^{-\xi y} \sin\xi x\, d\xi \\[5mm] \sigma_{xx} = -\sqrt{\dfrac{2}{\pi}} \displaystyle\int_0^\infty (1\text{-}\xi y)\, P(\xi)\, e^{-\xi y} \cos\xi x\, d\xi \\[5mm] \sigma_{yy} = -\sqrt{\dfrac{2}{\pi}} \displaystyle\int_0^\infty (1+\xi y)\, P(\xi)\, e^{-\xi y} \cos\xi x\, d\xi \end{array}\right\} \qquad (1.2.12)$$

where $P(\xi)$ is an unknown function to be determined. The constants μ and η are the rigidity modulus and the Poisson ratio, respectively, of the material of the elastic plate under consideration.

We find that the boundary conditions (1.2.11) give rise to the following relations for the determination of the unknown function $P(\xi)$:

$$\left.\begin{array}{l} \sqrt{\dfrac{2}{\pi}} \displaystyle\int_0^\infty P(\xi) \cos \xi x \, d\xi = p(x), \ 0 < x < 1 \\[4mm] \sqrt{\dfrac{2}{\pi}} \displaystyle\int_0^\infty \dfrac{P(\xi)}{\xi} \cos \xi x \, d\xi = 0, \ x > 1. \end{array}\right\} \tag{1.2.13}$$

It may be noted that the condition $\sigma_{xy} = 0$ on $y = 0$ is automatically satisfied and that

$$\frac{\mu}{1-\eta} u_y(x,0) = \sqrt{\frac{2}{\pi}} \int_0^\infty \frac{P(\xi)}{\xi} \cos\xi x \, d\xi \tag{1.2.14}$$

with the requirements that

$$u_y(x,0) = \begin{cases} 0(1) & \text{as } x \to 0, \\ 0\left((1-x^2)^{1/2}\right) & \text{as } x \to 1. \end{cases} \tag{1.2.15}$$

If we integrate the first equation in (1.2.13), we obtain

$$\sqrt{\frac{2}{\pi}} \int_0^\infty \frac{P(\xi)}{\xi} \sin\xi x \, d\xi = \int_0^x p(t) \, dt \equiv q(x), \ 0 < x < 1. \tag{1.2.16}$$

Setting

$$\sqrt{\frac{2}{\pi}} \int_0^\infty \frac{P(\xi)}{\xi} \cos\xi x \, d\xi = \varphi(x), \ 0 < x < 1, \tag{1.2.17}$$

and using the Fourier cosine inversion formula, after utilizing the second relation of (1.2.13), we get

$$\frac{P(\xi)}{\xi} = \sqrt{\frac{2}{\pi}} \int_0^1 \varphi(t) \cos\xi t \, dt. \tag{1.2.18}$$

Then the relations (1.2.16) and (1.2.18) give rise to the equation (see Chakrabarti (2008), p 113, for details)

$$\int_0^1 \frac{\varphi(t)}{t^2 - x^2} \, dt = -\frac{\pi}{2} \frac{q(x)}{x}, \quad 0 < x < 1 \tag{1.2.19}$$

where the integral is in the sense of CPV and $\varphi(t)$ satisfies the end conditions

$$\varphi(t) = \begin{cases} 0(1) & \text{as } t \to 0, \\ 0\left((1-t)^{1/2}\right) & \text{as } t \to 1. \end{cases} \tag{1.2.20}$$

Setting

$$\frac{\varphi(\sqrt{s})}{\sqrt{s}} = f(s), \quad -\frac{\pi q(\sqrt{u})}{\sqrt{u}} = g(u) \tag{1.2.21}$$

along with

$$t^2 = s, \ x^2 = u. \tag{1.2.22}$$

the equation (1.2.19) can be cast into the form

$$\int_0^1 \frac{f(s)}{s-u} \, ds = g(x), \ 0 < u < 1 \tag{1.2.23}$$

which is a singular integral equation with Cauchy-type kernel, and has to be solved under the conditions that

$$f(s) = \begin{cases} 0(s^{-1/2}) & \text{as } s \to 0, \\ 0\left((1-s)^{1/2}\right) & \text{as } s \to 1. \end{cases} \tag{1.2.24}$$

B. A mixed boundary value problem in the linearized theory of water waves

The problem of surface water wave scattering by a thin vertical barrier in the linearised theory of water waves is shown below to reduce to a homogeneous singular integral equation with Cauchy kernel. The mathematical formulation of the problem in this case is the following:

To solve the Laplace equation

$$\frac{\partial^2 \varphi}{\partial x^2} + \frac{\partial^2 \varphi}{\partial y^2} = 0, \ y > 0, \ -\infty < x < \infty \qquad (1.2.25)$$

with the boundary conditions

$$K\varphi + \frac{\partial \varphi}{\partial y} = 0 \ \text{ on } \ y = 0, \ -\infty < x < \infty \qquad (1.2.26)$$

where K is a positive real constant,

$$\frac{\partial \varphi}{\partial x} = 0 \ \text{ on } \ x = 0, \ 0 < y < a \qquad (1.2.27)$$

along with the following conditions, across x = 0:

$$\left.\begin{array}{l} \dfrac{\partial \varphi}{\partial x} \ \text{ is continuous on } \ x = 0, \ 0 < y < \infty \\[2mm] \varphi \ \text{ is continuous on } \ x = 0, \ a < y < \infty; \end{array}\right\} \qquad (1.2.28)$$

and the infinity conditions

$$\varphi(x, y) \rightarrow \begin{cases} e^{-Ky} \left(e^{iKx} + R\,e^{-iKx} \right) \ \text{ as } \ x \rightarrow -\infty \qquad (1.2.29) \\[2mm] T\,e^{-Ky + iKx} \ \text{ as } \ x \rightarrow \infty, \end{cases}$$

$$\varphi, \nabla\varphi \rightarrow 0 \ \text{ as } \ y \rightarrow \infty \qquad (1.2.30)$$

where R and T arc unknown physical constants called the reflection and transmission coefficients and $e^{-Ky + iKx}$ represents the incident field. The function $\varphi(x, y)$ also satisfies the edge condition

$$\frac{\partial \varphi}{\partial x} (0, y) = 0 \ (|\, y - a\,|^{-1/2}) \ \text{ as } \ y \rightarrow a. \qquad (1.2.31)$$

Setting

$$\varphi(x, y) = \varphi_0(x, y) + \psi(x, y) \qquad (1.2.32)$$

where

$$\varphi_0(x, y) = \begin{cases} e^{-Ky} \left(e^{iKx} + R\,e^{-iKx} \right) \ \text{ as } \ x < 0 \qquad (1.2.33) \\[2mm] T\,e^{-Ky + iKx}, \ x > 0, \end{cases}$$

and observing , because of the conditions at infinity, namely (1.2.29) and (1.2.30), that $\psi(x,y) \to 0$ as $|x| \to \infty$ and also as $y \to \infty$, we can have the representation of the unknown harmonic function $\psi(x,y)$ satisfying the boundary condition (1.2.26), as given by the following expressions:

$$\psi(x,y) = \begin{cases} \dfrac{2}{\pi} \displaystyle\int_0^\infty \dfrac{A(k)\,e^{-kx}}{k^2 + K^2} \, L(k,y)\,dk, \ x>0 \\[4mm] \dfrac{2}{\pi} \displaystyle\int_0^\infty \dfrac{B(k)\,e^{kx}}{k^2 + K^2} \, L(k,y)\,dk, \ x<0 \end{cases} \tag{1.2.34}$$

where $A(k)$ and $B(k)$ are two unknown functions to be determined, and

$$L(k,y) = k \, \cos ky - K \, \sin ky. \tag{1.2.35}$$

Using the Fourier analysis, we obtain the pair of formulae, as given by

$$F(k) = \int_0^\infty f(y)\, L(k,y)\, dy, \tag{1.2.36}$$

$$f(y) = 2K\, e^{-Ky} \int_0^\infty f(t)\, e^{-Kt} dt + \frac{2}{\pi} \int_0^\infty \frac{F(k)}{k^2 + K^2}\, L(k,y)\, dk, \ y>0.$$

Now, using the representations (1.2.34) for the function $\psi(x,y)$ along with (1.2.32) and (1.2.33), we find that the continuity conditions (1.2.25) give rise to the following important relations:

$$\frac{2}{\pi} \int_0^\infty \frac{A(k) - B(k)}{k^2 + K^2}\, L(k,y)\, dk = (1+R-T)\, e^{-Ky}, \ a<y<\infty, \tag{1.2.37}$$

$$\frac{2}{\pi} \int_0^\infty \frac{k\{A(k) + B(k)\}}{k^2 + K^2}\, L(k,y)\, dk = -iK(1-R-T)e^{-Ky}, \ a<y<\infty, \tag{1.2.38}$$

$$\frac{2}{\pi} \int_0^\infty \frac{kA(k)}{k^2 + K^2}\, L(k,y)\, dk = -iKT\, e^{-Ky}, \ 0<y<a, \tag{1.2.39}$$

$$\frac{2}{\pi} \int_0^\infty \frac{kB(k)}{k^2 + K^2}\, L(k,y)\, dk = -iK\,(1-R)\, e^{-Ky}, \ 0<y<a. \tag{1.2.40}$$

Adding both sides of the relations (1.2.39) and (1.2.40) we find that

$$\frac{2}{\pi} \int_0^\infty \frac{k\{A(k)+B(k)\}}{k^2+K^2} L(k,y)\, dk \;=\; -iK\,(1-R-T)\, e^{-Ky},\; a<y<\infty. \quad (1.2.41)$$

Using the relations (1.2.35) and (1.2.41) along with the pair (1.2.36), we find that we must have

$$A(k) = -B(k), \quad\quad\quad (1.2.42)$$

$$R+T = 1. \quad\quad\quad (1.2.43)$$

We thus obtain from (1.2.39) and (1.2.37)

$$\frac{2}{\pi} \int_0^\infty \frac{kA(k)}{k^2+K^2} L(k,y)\, dk \;=\; K(1-R)\, e^{-Ky},\; 0<y<a, \quad (1.2.44)$$

$$\frac{2}{\pi} \int_0^\infty \frac{A(k)}{k^2+K^2} L(k,y)\, dk = 2R\, e^{-Ky},\; a<y<\infty. \quad (1.2.45)$$

Setting

$$\frac{2}{\pi} \int_0^\infty \frac{kA(k)}{k^2+K^2} L(k,y)\, dk = h(y),\; a<y<\infty, \quad (1.2.46)$$

where $h(y)\ (y>a)$ is an unknown function to be determined, we find with the aid of (1.2.36)

$$kA(k) = iK(1-R)\, e^{-Ka} \sin ka + \int_a^\infty h(t)\, L(k,t)\, dt \quad (1.2.47)$$

with

$$\int_a^\infty h(t)\, e^{-Kt} dt = -\frac{i}{2}\,(1-R)\,(1-e^{-2Ka}). \quad (1.2.48)$$

Substituting $A(k)$ from (1.2.47) into the relation (1.2.45) we find that the function $h(t)$ must satisfy the equation

$$\frac{2}{\pi} \int_0^\infty \frac{L(k,y)}{k(k^2+K^2)} \left\{ \int_a^\infty h(t)\, L(k,t)\, dt \right\} dk$$

$$+\frac{2i}{\pi} K(1-R)\, e^{-Ka} \int_0^\infty \frac{\sin ka\; L(k,y)}{k(k^2+K^2)}\, dk = 2R\, e^{-Ky},\; y>a. \quad (1.2.49)$$

Operating on both sides of the equation (1.2.49) by $\dfrac{d}{dy} + K$ along with the results

$$\int_0^\infty \frac{\sin ky \, \sin kt}{k} \, dk = \frac{1}{2} \, \ln \left| \frac{y-t}{y+t} \right|, \quad t, y > 0,$$

$$\lim_{\varepsilon \to +0} \int_0^\infty e^{-\varepsilon K} \, \sin ky \, \sin kt \, dk = \frac{y}{y^2 - t^2}, \quad t, y > 0,$$

we obtain the equation

$$\int_a^\infty h(t) \left(\frac{1}{y+t} + \frac{1}{y-t} + K \, \ln \left| \frac{y-t}{y+t} \right| \right) dt = iK \, e^{-Ka} \, (1-R) \, \ln \left| \frac{y-a}{y+a} \right|, \quad y > a. \quad (1.2.50)$$

If we use the result

$$\int_a^\infty e^{-Kt} \left(\frac{1}{y+t} + \frac{1}{y-t} + K \, \ln \left| \frac{y-t}{y+t} \right| \right) dt = e^{-Ka} \, \ln \left| \frac{y-a}{y+a} \right|, \quad y > a, \quad (1.2.51)$$

we find that the equation (1.2.50) becomes

$$\int_a^\infty g(t) \left(\frac{1}{y+t} + \frac{1}{y-t} + K \, \ln \left| \frac{y-t}{y+t} \right| \right) dt = 0, \quad y > a, \quad (1.2.52)$$

where

$$g(t) = h(t) - iK(1-R) \, e^{-Kt}, \quad t > a. \quad (1.2.53)$$

Now, from the relations (1.2.32) and (1.2.33) we find that

$$\frac{\partial \varphi}{\partial x} (+0, y) = iKT \, e^{-Ky} + \frac{\partial \psi}{\partial x} (+0, y)$$

$$= iK(1-R) \, e^{-Ky} - \frac{2}{\pi} \int_0^\infty \frac{kA(k)}{k^2 + K^2} \, L(k, y) \, dk \quad (1.2.54)$$

$$= iK(1-R) \, e^{-Ky} - h(y), \quad y > a.$$

Thus by (1.2.54) and (1.2.53) along with the edge condition (1.2.31), we observe that the physical problem under consideration reduces to that of solving a homogeneous singular integral equation (1.2.52), for the unknown function $g(y)$ which must satisfy the condition at the end point $y = a$ given by

$$g(y) = 0(\mid y - a \mid^{-1/2}) \quad \text{as} \quad y \to a. \quad (1.2.55)$$

We also note that

$$\lim_{y \to \infty} g(y) = 0. \tag{1.2.56}$$

After determining the function $g(y)$, i.e., $h(y)$ completely, we will be in a position to determine the unknown physical constant R (the reflection coefficient), by using the relation (1.2.48). The details of these calculations constitute a different study altogether, and will not be considered here.

For the present moment, we will reduce the singular integral equation (1.2.53) to a Cauchy singular integral equation. For this we set

$$G(t) = \int_a^t g(s) \, ds \tag{1.2.57}$$

so that

$$G'(t) = g(t) \tag{1.2.58}$$

and

$$G(a) = 0. \tag{1.2.59}$$

We now observe that

$$\int_a^\infty g(t) \, \ln \left| \frac{y-t}{y+t} \right| \, dt = \int_a^\infty G(t) \left(\frac{1}{y+t} + \frac{1}{y-t} \right) dt.$$

Thus the equation (1.2.52) can be reduced to the form

$$\int_a^\infty p(t) \left(\frac{1}{y+t} + \frac{1}{y-t} \right) dt = 0, \quad y > a$$

which is equivalent to

$$\int_a^\infty \frac{p(t)}{t^2 - y^2} \, dt = 0, \quad y > a \tag{1.2.60}$$

where

$$p(t) = g(t) + K\, G(t)$$

and the integral is in the sense of CPV. The equation (1.2.60) is equivalent to the homogeneous singular integral equation

$$\int_0^1 \frac{q(u)}{u - v} \, du = 0, \quad 0 < v < 1 \tag{1.2.61}$$

where

$$q(u) = \frac{p(au^{-1/2})}{u^{1/2}}. \tag{1.2.62}$$

The equation (1.2.61) has to be solved under the end conditions

$$q(u) = \begin{cases} 0(u^{-1/2}) & \text{as } u \to 0, \\ 0\left((1-u)^{-1/2}\right) & \text{as } u \to 1. \end{cases} \tag{1.2.63}$$

1.2.3 Hypersingular integral equation

Many two-dimensional boundary value problems involving thin obstacles can be reduced to hypersingular integral equations. Martin (1991) gave a number of examples from potential theory, acoustics, hydrodynamics and elastostatics. A simple example involving two-dimensional flow past a rigid plate in an infinite fluid.

Let $\varphi_0(x, y)$ be the known potential describing the two-dimensional flow in an infinite fluid in the absence of a rigid plate occupying the position $y = 0$, $0 < x < a$. Let $\varphi(x, y)$ be the potential due to the presence of the rigid plate so that the total potential is

$$\varphi^{tot}(x, y) = \varphi_0(x, y) + \varphi(x, y). \tag{1.2.64}$$

The function $\varphi(x, y)$ satisfies the Laplace equation

$$\nabla^2 \varphi = 0 \quad \text{in the fluid region}, \tag{1.2.65}$$

the condition on the plate

$$\frac{\partial \varphi}{\partial y} = -\frac{\partial \varphi_0}{\partial y} \quad \text{on } y = 0, \ 0 < x < a, \tag{1.2.66}$$

the edge conditions

$$\varphi = 0(1) \quad \text{near } (0,0) \text{ and } (0,a) \tag{1.2.67}$$

and the infinity condition

$$\varphi \to 0 \quad \text{as } r = \left(x^2 + y^2\right)^{1/2} \to \infty. \tag{1.2.68}$$

Let

$$G(x, y; \xi, \eta) = \frac{1}{2} \, ln \, \{(x-\xi)^2 + (y-\eta)^2\}. \qquad (1.2.69)$$

We apply Green's theorem to $\varphi(x, y)$ and $G(x, y; \xi, \eta)$ in the region bounded externally by a circle of *large* radius R with centre at the origin, and internally by a circle of *small* radius ε with centre at (ξ, η) and a contour enclosing the plate, and ultimately make $R \to \infty$, $\varepsilon \to 0$ and the contour around the plate to shrink into it. We then obtain

$$\varphi(\xi, \eta) = -\frac{\eta}{2\pi} \int_0^a \frac{f(x)}{(x-\xi)^2 + (y-\eta)^2} \, dx \qquad (1.2.70)$$

where

$$f(x) = \varphi(x, +0) - \varphi(x, -0), \; 0 < x < a \qquad (1.2.71)$$

and is unknown. $f(x)$ satisfies the end conditions

$$f(0) = 0, \; f(a) = 0. \qquad (1.2.72)$$

$f(x)$ can be found by using the condition (1.2.66) on the plate written in terms of ξ, η i.e.

$$\frac{\partial \varphi}{\partial \eta}(\xi, 0) = -\frac{\partial \varphi_0}{\partial \eta}(\xi, \eta) \equiv -\frac{1}{2\pi} v(\xi), \; 0 < \xi < a \qquad (1.2.73)$$

where $v(\xi)$ is a known function. Using the representation (1.2.70) we obtain

$$\times \int_0^a \frac{f(x)}{(x-\xi)^2} \, dx = v(\xi), \; 0 < \xi < a, \qquad (1.2.74)$$

where the integral is in the sense of Hadamard finite part of order 2. The equation (1.2.74) is the simplest hypersingular integral equation.

Chapter 2

Some Elementary Methods of Solution of Singular Integral Equations

In this chapter we present some elementary methods to solve certain singular integral equations of some special types and classes. As applications of such elementary methods of solutions we take up the integral equations arising in some problems of in the theory of elasticity and surface water wave scattering.

2.1 ABEL INTEGRAL EQUATION AND ITS GENERALIZATION

In this section we present some Abel type integral equations and their solutions

(a) We first consider the Abel integral equation as given by

$$(\tilde{A}\varphi)\ \varphi(x) = \frac{1}{\pi}\ \frac{d}{dx}\left[\int_0^x \frac{f(t)}{(x-t)^{1/2}}\ dt\right].\tag{2.1.1}$$

with $f(0) = 0$, where the operator $\tilde{A}$ may be regarded as the Abel integral operator. As mentioned in Chapter 1, this integral equation was discovered by Abel in 1826 and is the first equation in the theory of integral equation.

We can solve the integral equation (2.1.1) for the class of functions whose Laplace transforms exist. The *Laplace transform* of the functions $\varphi(x)$ and $f(x)$ are defined by (cf. Doetsch (1955), Sneddon (1974))

$$\big(\Phi(p), F(p)\big) = \int_0^\infty \big(\varphi(x), f(x)\big)\ \mathrm{e}^{-px}\ dx,\quad Re\ p > \delta > 0\tag{2.1.2}$$

where δ is some positive number. The *inverse formulae* for the Laplace transforms are

$$(\varphi(x), f(x)) = \frac{1}{2\pi i} \int_{\gamma - i\infty}^{\gamma + i\infty} (\Phi(p), F(p)) \, e^{px} \, dp, \; x > 0 \qquad (2.1.3)$$

where γ is greater than the real part of the singularities of $\Phi(p)$ and $F(p)$. It may be noted that γ may be different for $\Phi(p)$ and $F(p)$.

The *convolution theorem* involving Laplace transforms $F(p)$ and $K(p)$ of $f(x)$ and $k(x)$ respectively is

$$\int_0^\infty \left\{ \int_0^x f(t) \, k(x-t) dt \right\} e^{-px} \, dx = F(p) \, K(p) \qquad (2.1.4)$$

for $Re \, p > \delta_1 > 0$ where δ_1 is some positive number, and for some special classes of functions $f(x)$ and $k(x)$. The details are available in the treatise by Doetsch (1955) and Sneddon (1974). Then using Laplace transform to the both sides of equation (2.1.1), along with the convolution theorem (2.1.4), we find that

$$\Phi(p) \, K(p) = F(p) \qquad (2.1.5)$$

where

$$K(p) = \int_0^\infty k(x) \, e^{-px} \, dx \quad \left(k(x) = x^{-1/2}\right)$$

$$\qquad (2.1.6)$$

$$= \sqrt{\frac{\pi}{p}} \quad \text{for } Re \, p > 0.$$

The relation (2.1.5) can thus be expressed as

$$\Phi(p) = \frac{p}{\pi} \sqrt{\frac{\pi}{p}} \, F(p) \qquad (2.1.7)$$

which is assumed to hold good for $Re \, p > \delta > 0$ where δ is some positive number and depends on the class of functions φ and f.

If we next use the well-known result involving Laplace transforms, as given by,

$$\int_0^\infty h'(x) \, e^{-px} \, dx = p \, H(p) - h(0) \qquad (2.1.8)$$

where $h'(x)$ denotes the derivative of $h(x)$, $H(p)$ is the Laplace transform of $h(x)$, and also the convolution theorem (2.1.4), we find from the relation (2.1.7) that the solution function $\varphi(x)$ can be expressed in either of the two forms:

$$\varphi(x) = \frac{1}{\pi} \int_0^x \frac{f'(t)}{(x-t)^{1/2}} \, dt \qquad (2.1.9a)$$

where $f'(t)$ denotes the derivative of $f(t)$, since $f(0) = 0$, after utilizing the result (2.1.6),

$$\varphi(x) = \frac{1}{\pi} \frac{d}{dx} \left[\int_0^x \frac{f(t)}{(x-t)^{1/2}} \, dt \right]. \qquad (2.1.9b)$$

Remarks

1. The formulae (2.1.9a) and (2.1.9b) represent two different forms of the solution of the Abel integral equation (2.1.1).

2. The formula (2.1.9b) is known as the general inversion formula for the Abel operator $\tilde{A}$, defined in the equation (2.1.1), and the formula (2.1.9a) is a special case of the formula (2.1.9b), in the circumstances when $f(x)$ is a differentiable function with $f(0) = 0$. In fact, the formula (2.1.9a) can be derived from formula (2.1.9b), by an integration by parts whenever $f(x)$ is differentiable and $f(0) = 0$.

3. Another *very elementary* method to solve the integral equation (2.1.1) is to multiply both sides by $(y-x)^{-1/2}$ and integrate w.r.to x between 0 to y. This produces

$$\int_0^y \frac{f(x)}{(y-x)^{1/2}} \, dx = \int_0^y \left\{ \int_0^x \frac{\varphi(t)}{(x-t)^{1/2}} \, dt \right\} \frac{dx}{(y-x)^{1/2}}$$

$$= \int_0^y \left\{ \int_t^y \frac{dx}{(x-t)^{1/2}(y-x)^{1/2}} \right\} \varphi(t) \, dt$$

$$= \pi \int_0^y \varphi(t) \, dt.$$

Then by differentiation w.r.to y, we obtain

$$\varphi(y) = \frac{1}{\pi} \frac{d}{dy} \left[\int_0^y \frac{f(x)}{(y-x)^{1/2}} \, dx \right]$$

which is the same as (2.1.9b).

(b) The integral equation (2.1.1) is sometimes called Abel integral equation of the *first type*. The *second type* Abel integral equation is

$$\int_x^b \frac{\varphi(t)}{(t-x)^{1/2}}\, dt = f(x),\ 0 < x < b \tag{2.1.10}$$

where $f(b) = 0$. Its solution can easily be obtained as

$$\varphi(x) = -\frac{1}{\pi}\frac{d}{dx}\int_x^b \frac{f(t)}{(t-x)^{1/2}}\, dt. \tag{2.1.11}$$

(c) A slight generalization of *first type* Abel integral equation is

$$\int_0^x \frac{\varphi(t)}{(x-t)^{\alpha}}\, dt = f(x),\ x > 0, \tag{2.1.12}$$

where $f(0) = 0$ and $0 < \alpha < 1$. Its solution can be obtained, using the Laplace transform method or an obvious very elementary method, as

$$\varphi(x) = \frac{\sin\pi\alpha}{\pi}\frac{d}{dx}\left[\int_0^x \frac{f(t)}{(x-t)^{1-\alpha}}\, dt\right],\ x > 0. \tag{2.1.13}$$

(d) The solution of the *second type* Abel integral equation

$$\int_x^b \frac{\varphi(t)}{(t-x)^{\alpha}}\, dt = f(x),\ 0 < x < b,$$

where $f(b) = 0,\ 0 < \alpha < 1$, is

$$\varphi(x) = -\frac{\sin\pi\alpha}{\pi}\frac{d}{dx}\left[\int_x^b \frac{f(t)}{(t-x)^{1-\alpha}}\, dt\right],\ 0 < x < b. \tag{2.1.14}$$

(e) The most *general form* of *first type* Abel integral equation is

$$\int_a^x \frac{\varphi(t)}{\{h(x)-h(t)\}^{\alpha}}\, dt = f(x),\ a < x < b, \tag{2.1.15}$$

where $f(a) = 0,\ 0 < \alpha < 1$ and $h(x)$ is a strictly monotonically increasing and differentiable function of x on $[a,b]$ and $h'(x) \neq 0$ on $[a,b]$.

Its solution is

$$\varphi(x) = \frac{\sin \pi \alpha}{\pi} \frac{d}{dx} \left[\int_a^x \frac{f(t)\, h'(t)}{\{h(x) - h(t)\}^{1-\alpha}}\, dt \right], \quad a < x < b. \quad (2.1.16)$$

(f) The *most general form* of *second type* Abel integral equation is

$$\int_x^b \frac{\varphi(t)}{\{h(t) - h(x)\}^{1/2}}\, dt = f(x), \quad a < x < b, \quad (2.1.17)$$

where $f(b) = 0$, $0 < \alpha < 1$ and $h(x)$ is as in (e) above.

Its solution is

$$\varphi(x) = -\frac{\sin \pi \alpha}{\pi} \frac{d}{dx} \left[\int_x^b \frac{f(t)\, h'(t)}{\{h(t) - h(x)\}^{1-\alpha}}\, dt \right], \quad a < x < b. \quad (2.1.18)$$

(g) *A special case*

For the special case $h(x) = x^2$, $a = 0$, $b = 1$ and $\alpha = \dfrac{1}{2}$, the Abel integral equation of first kind (2.1.15) has the form

$$(A\varphi)(x) \equiv \int_0^x \frac{\varphi(t)}{\left(x^2 - t^2\right)^{1/2}}\, dt = f(x),\ 0 < x < 1\ (f(0) = 0) \quad (2.1.19)$$

having the solution

$$\varphi(x) = (A^{-1} f)(x) = \frac{2}{\pi} \frac{d}{dx} \left[\int_0^x \frac{t\, f(t)}{\left(x^2 - t^2\right)^{1/2}}\, dt \right],\ 0 < x < 1 \quad (2.1.20)$$

while the Abel integral equation of the second kind (2.1.17) has the form

$$\int_x^1 \frac{\varphi(t)}{(t^2 - x^2)^{1/2}}\, dt = f(x),\ 0 < x < 1\ \ (f(1) = 0) \quad (2.1.21)$$

having the solution

$$\varphi(x) = -\frac{2}{\pi} \frac{d}{dx} \left[\int_x^1 \frac{t\, f(t)}{\left(t^2 - x^2\right)^{1/2}}\, dt \right],\ 0 < x < 1. \quad (2.1.22)$$

(h) *More general Abel type integral equation*

If we now introduce the operators B and D, as defined by

$$(Bf)(x) = \frac{1}{\pi} \frac{d}{dx} f(x)$$

$$(Df)(x) = 2x f(x) \tag{2.1.23}$$

we find that the inverse operator A^{-1}, as given by the relation (2.1.20), can be expressed as

$$(A^{-1}f)(x) = [(BAD)f](x). \tag{2.1.24}$$

This way of expressing the inverse operator A^{-1}, has been mentioned by Knill (1994), who has demonstrated a method, known as the *diagonalization method* for solving the Abel type integral equation (2.1.19), Chakrabarti and George (1997) have generalized the idea of Knill (1994) further and have explained the diagonalization method for more general Abel type integral equations as given by

$$\int_0^x \frac{k(x,t)}{\left(x^\beta - t^\beta\right)^\alpha} \varphi(t)\, dt = f(x), \ x > 0 \tag{2.1.25}$$

where $0 < \alpha < 1, \ \beta > 0$ and

$$k(x,t) = \sum_{j=1}^m a_j \, x^{\alpha\beta - j} \, t^{j-1}, \tag{2.1.26}$$

$a_j \ (j = 1,2,...m)$ being known constants.

It has been shown by Chakrabarti and George (1997) that, the solution of the integral equation (2.1.25) can be expressed in the form

$$\varphi(x) = \sum_{n=0}^\infty \frac{f_n}{\mu_n} x^n \ \ for \ \ 0 \le x \le r, \tag{2.1.27}$$

for all $f(x)$ such that

$$f(x) = \sum_{n=0}^\infty f_n x^n \ \ for \ \ 0 \le x \le r \tag{2.1.28}$$

and

$$\mu_n = \frac{\Gamma(1-\alpha)}{\beta} \sum_{j=1}^{m} a_j \frac{\Gamma\left(\dfrac{n+j}{\beta}\right)}{\Gamma\left(\dfrac{n+j}{\beta}+1-\alpha\right)}. \qquad (2.1.29)$$

Details are omitted here and the reader is referred to the work of Chakrabarti and George (1997) for details.

(i)   *An important result*

Let

$$\int_x^b \frac{h_1'(t)\varphi(t)}{\{h_1(t)-h_1(x)\}^{1/2}} dt = \int_x^b \frac{h_2'(t)\,\psi(t)}{\{h_2(t)-h_2(x)\}^{1/2}} dt, \ 0<x<b \qquad (2.1.30)$$

where

$\varphi(b) = 0, \psi(b) = 0$; $h_1(t)$, $h_2(t)$ are monotonically increasing functions in $(0,b)$; $h_1(0) = 0$; $h_2(0) = 0$; $h_1(t)$ and $h_2(t)$ are even functions of t. Then

$$\int_0^x \frac{h_1'(t)\,\varphi(t)}{\{h_1(x)-h_1(t)\}^{1/2}} dt = \int_0^x \frac{h_2'(t)\,\psi(t)}{\{h_2(x)-h_2(t)\}^{1/2}} dt, \ 0<x<b. \qquad (2.1.31)$$

This result has been proved in the paper of De, Mandal and Chakrabarti (2009). It has been successfully utilized in the study of water wave scattering problems involving two submerged plane vertical barriers and two surface piercing barriers (De et al. (2009, 2010)). This result is also true for $a < x < b$, and in that case the lower limit of the integrals in both sides of (2.1.31) is a.

2.2 INTEGRAL EQUATIONS WITH LOGARITHMIC TYPE SINGULARITIES

In this section we present some elementary methods of solution of weakly singular equations with logarithmic type singularities.

(a) *Reduction to a singular integral equation of Cauchy type*

The integral equation with logarithmic type singularity, as given by

$$\int_a^b \ln |x-t| \, \varphi(t) \, dt = f(x), \ a < x < b, \qquad (2.2.1)$$

can be solved, for some specific class of functions $\varphi(x), f(x)$, by differentiating the integral equation (2.2.1) with respect to x and solving the resulting singular integral equation of the Cauchy type, as given by

$$\int_a^b \frac{\varphi(t)}{x-t} \, dt = f'(x), \ a < x < b \qquad (2.2.2)$$

where the integral is in the sense of Cauchy principle value. The domain (a,b) of the integral equation (2.2.2) can be transformed into the interval $(0,1)$ by using the transformations

$$t = \frac{u-a}{b-a}, \ x = \frac{v-a}{b-a}$$

with u, v being the new variables, and then the solution of the transformed integral equation

$$\int_0^1 \frac{\psi(u)}{v-u} \, du = h(v), \ 0 < v < 1 \qquad (2.2.3)$$

where

$$\psi(u) = \frac{1}{b-a} \, \varphi\left(\frac{u-a}{b-a}\right), \ h(v) = f'\left(\frac{v-a}{b-a}\right), \qquad (2.2.4)$$

will finally solve the integral equation (2.2.2) completely.

In section 2.3 of Chapter 2, we will present a simple method of solution of the integral equation (2.2.2).

(b) *Reduction to a Riemann-Hilbert problem*

Here a method of solution of the singular integral equation, with a logarithmic singularity, as given by

$$\frac{1}{\pi} \int_0^1 \varphi(t) \ln \left| \frac{x+t}{x-t} \right| \, dt = f(x), \ 0 < x < 1 \qquad (2.2.5)$$

is explained briefly. In the equation (2.2.5), φ and f are assumed to be differentiable in $(0,1)$.

We first extend the integral equation (2.2.5) into the extended interval $(-1,1)$, by using it as

$$\frac{1}{\pi} \int_0^1 \varphi(t) \ln \left| \frac{x+t}{x-t} \right| dt = F(x), \quad -1 < x < 1 \qquad (2.2.6)$$

where

$$F(x) = \begin{cases} f(x), & 0 < x < 1, \\ -f(-x), & -1 < x < 0. \end{cases} \qquad (2.2.7)$$

Then if we set

$$\Phi(z) = \frac{d}{dz} \int_0^1 \varphi(t) \ln \left(\frac{t+z}{t-z} \right) dt, \qquad (2.2.8)$$

we obtain a sectionally analaytic function $\Phi(z)$, which is analytic in the complex z-plane ($z = x + iy$, $i^2 = -1$) cut along the real axis from $z = -1$ to $z = 1$, along with the following properties

(i)
$$\Phi(z) = \frac{d}{dz} \left[\int_0^x \varphi(t) \{ \ln (t+z) - \ln (t-z) \} dt \right.$$

$$\left. + \int_x^1 \varphi(t) \{ \ln (t+z) - \ln (t-z) \} dt \right]$$

(ii)
$$\lim_{z \to x \pm i0} \Phi(z) \equiv \Phi^{\pm}(x) = \frac{d}{dx} \left[\int_0^1 \varphi(t) \ln \left| \frac{t+x}{t-x} \right| dt \right] \mp i\pi\varphi(x)$$

and

(iii)
$$\Phi(z) = 0 \left(\frac{1}{z^2} \right) \text{ for large } |z| \text{ so that } \lim_{z \to \infty} \Phi(z) = 0$$

Using the limiting relations (ii), the integral equation (2.2.6) can be expressed as a functional relation as given by

$$\Phi^+(x) + \Phi^-(x) = g(x), \quad -1 < x < 1 \qquad (2.2.9)$$

where

$$g(x) = \begin{cases} 2\pi \, f'(x), & 0 < x < 1, \\ 2\pi \, f'(-x), & -1 < x < 0 \end{cases} \qquad (2.2.10)$$

with dash denoting differentiation with respect to the argument.

The functional relation (2.2.9) is one of the types that arises in a more general problem, called the *Riemann-Hilbert Problem* (RHP), which will be described in some detail in Chapter 3.

The general solution of the RHP (2.2.9) can be expressed as

$$\Phi(z) = \frac{1}{z(z^2-1)^{1/2}} \left[\frac{1}{2\pi i} \int_{-1}^{1} \frac{g(t)}{\Phi_0^+(t)} \frac{dt}{t-z} + D \right] \quad (2.2.11)$$

with

$$\Phi_0(z) = \frac{1}{z\left(z^2-1\right)^{1/2}} \quad (2.2.12)$$

being the solution of the *homogeneous* RHP (2.2.4), giving

$$\Phi_0^{\pm}(t) = \mp \frac{i}{t(1-t^2)^{1/2}}, \quad (-1 < t < 1), \quad (2.2.13)$$

and D being an arbitrary constant. Solution of the integral equation (2.2.5) is obtained by using the relation (ii), in the form

$$\varphi(x) = \frac{1}{2\pi i} \left[\Phi^+(x) - \Phi^-(x) \right] = \frac{2}{\pi x\left(1-x^2\right)^{1/2}} \left[\int_{0}^{1} \frac{t^2\left(1-t^2\right)^{1/2} f'(t)}{x^2-t^2} dt + D_0 \right] \quad (2.2.14)$$

where D_0 is an arbitrary constant.

2.3 INTEGRAL EQUATIONS WITH CAUCHY TYPE KERNELS

In this section we consider the problem of determining solutions of Cauchy type singular integral equations using elementary methods.

(a) *First kind singular integral equation with Cauchy kernel*

We consider the simple first kind Cauchy type singular integral equation

$$\frac{1}{\pi} \int_{a}^{b} \frac{\varphi(t)}{x-t} dt = f(x), \quad a < x < b \quad (2.3.1)$$

where the integral is in the sense of Cauchy principal value, $f(x)$ is a known continuous function in (a,b). Its solution depends on the behaviour of $\varphi(t)$ at the end points dictated by the physics of the problem in which the integral equation arises. The solutions can be obtained for the following three forms of end behaviours:

(i) $\varphi(t) = 0(|t-a|^{-1/2})$ as $t \to a$ and $\varphi(t) = 0(|t-b|^{-1/2})$ as $t \to b$;

(ii) $\varphi(t) = 0(|t-a|^{-1/2})$ as $t \to a$ and $\varphi(t) = 0(|t-b|^{1/2})$ as $t \to b$,

   or, $\phi(t) = 0(|t-a|^{1/2})$ as $t \to a$ and $\phi(t) = 0(|t-b|^{-1/2})$ as $t \to b$;

(iii) $\varphi(t) = 0(|t-a|^{1/2})$ as $t \to a$ and $\varphi(t) = 0(|t-b|^{1/2})$ as $t \to b$;

For the form (i), the solution involves an arbitrary constant while for the form (ii), the solution does not involve any arbitrary constant. For the third form the solution exists if and only if $f(x)$ satisfies a certain condition known as the *solvability criterion* in the literature.

Here we use an elementary method to solve the integral equation (2.3.1). As in section 2.2(a), the domain (a,b) of the integral equation (2.3.1) can be transformed into the interval $(0,1)$. Thus, without any loss of generality, we consider the solution of

$$\frac{1}{\pi} \int_0^1 \frac{\varphi(t)}{x-t} \, dt = f(x), \ 0 < x < 1 \qquad (2.3.2)$$

where the integral is in the sense of CPV.

We now put $t = \xi^2$, $x = \eta^2$ $(\xi > 0, \eta > 0)$ in the equation (2.3.2) to obtain

$$\frac{1}{\pi} \int_0^1 \psi(\xi) \left(\frac{1}{\eta - \xi} - \frac{1}{\eta + \xi} \right) d\xi = f_1(\eta), \ 0 < \eta < 1 \qquad (2.3.3)$$

where

$$\psi(\xi) = \varphi(\xi^2), \ f_1(\eta) = f(\eta^2). \qquad (2.3.4)$$

Integrating both sides of (2.3.3) with respect to η between 0 to η we obtain

$$\frac{1}{\pi} \int_0^1 \psi(\xi) \ ln \left| \frac{\eta - \xi}{\eta + \xi} \right| d\xi = g(\eta), \ 0 < \eta < 1 \qquad (2.3.5)$$

where

$$g(\eta) = \int_0^\eta f_1(s) \, ds. \qquad (2.3.6)$$

We use the integral identity

$$\ln\left|\frac{\xi-\eta}{\xi+\eta}\right| = -2 \int_0^{\min(\xi,\eta)} \frac{u}{\left(\eta^2-u^2\right)^{1/2}\left(\xi^2-u^2\right)^{1/2}} \, du$$

to obtain from (2.3.5), after interchange of order of integration,

$$\int_0^{\eta}\left(\int_u^1 \frac{\psi(\xi)}{\left(\xi^2-u^2\right)^{1/2}} \, d\xi\right) \frac{u}{\left(\eta^2-u^2\right)^{1/2}} \, du = -\frac{\pi}{2}\, g(\eta), \quad 0<\eta<1. \quad (2.3.7)$$

This is equivalent to the pair of Abel integral equations

$$\int_u^1 \frac{\psi(\xi)}{\left(\xi^2-u^2\right)^{1/2}} \, d\xi = F(u), \quad 0<u<1, \quad\quad (2.3.8a)$$

$$\int_0^{\eta} \frac{u\, F(u)}{\left(\eta^2-u^2\right)^{1/2}} \, du = -\frac{\pi}{2}\, g(\eta), \quad 0<\eta<1. \quad\quad (2.3.8b)$$

Solution of the Abel integral equation (2.3.8b) follows from (2.1.20) and is given by

$$uF(u) = -\frac{d}{du}\left[\int_0^u \frac{\eta\, g(\eta)}{\left(u^2-\eta^2\right)^{1/2}} \, d\eta\right]$$

so that

$$= u \int_0^u \frac{f_1(\eta)}{\left(u^2-\eta^2\right)^{1/2}} \, d\eta$$

$$F(u) = \int_0^u \frac{f_1(\eta)}{\left(u^2-\eta^2\right)^{1/2}} \, d\eta. \quad\quad (2.3.9)$$

Again, solution of the Abel integral equation (2.3.8a) follows from (2.1.22) and is given by

$$\psi(u) = -\frac{2}{\pi}\frac{d}{du}\left[\int_u^1 \frac{\eta\, F(\eta)}{\left(\eta^2-u^2\right)^{1/2}} \, d\eta\right]$$

$$= -\frac{2}{\pi}\frac{u}{\left(1-u^2\right)^{1/2}} \int_0^1 \frac{\left(1-t^2\right)^{1/2} f_1(t)}{\left(u^2-t^2\right)} \, dt \quad\quad (2.3.10)$$

where the integral is in the sense of CPV. Back substitution of $u = y^{1/2}$, $t = x^{1/2}$ produces

$$\varphi(y) = -\frac{1}{\pi} \left(\frac{y}{1-y}\right)^{1/2} \int_0^1 \left(\frac{1-x}{x}\right)^{1/2} \frac{f(x)}{y-x} \, dx. \qquad (2.3.11)$$

The form (2.3.11) of the solution satisfies the end conditions

$$\varphi(x) = \begin{cases} 0\left(|x|^{1/2}\right) & \text{as } x \to 0 \\ 0\left(1-x|^{-1/2}\right) & \text{as } x \to 1. \end{cases}$$

From this solution we can derive in a *non-rigorous* manner, the solution for the case when

$$\varphi(x) = 0\left(|x|^{-1/2}\right) \text{ as } x \to 0 \text{ and } \varphi(x) = 0\left(1-x|^{-1/2}\right) \text{ as } x \to 1.$$

Using (2.3.11) we find that

$$\int_0^1 \varphi(y) \, dy = -\int_0^1 \left(\frac{1-x}{x}\right)^{1/2} f(x) \left[\frac{1}{\pi} \int_0^1 \left(\frac{y}{1-y}\right)^{1/2} \frac{dy}{y-x}\right] dx.$$

The integral in the square bracket has the value π so that

$$\int_0^1 \varphi(x) \, dx = -\int_0^1 \left(\frac{1-t}{t}\right)^{1/2} f(t) \, dt.$$

From the form (2.3.11) we find that

$$\varphi(y) = -\frac{1}{\pi} \frac{1}{\{y(1-y)\}^{1/2}} \left[-\int_0^1 \varphi(x) \, dx + \int_0^1 \{t(1-t)\}^{1/2} \frac{f(t)}{y-t} \, dt\right].$$

Writing $C = -\int_0^1 \varphi(x) \, dx$ we find that

$$\varphi(y) = -\frac{1}{\pi} \frac{1}{\{y(1-y)\}^{1/2}} \left[C + \int_0^1 \{t(1-t)\}^{1/2} \frac{f(t)}{y-t} \, dt\right]. \quad (2.3.12)$$

In view of the result that

$$\int_0^1 \frac{1}{\{y(1-y)\}^{1/2}} \frac{dt}{y-t} = 0 \quad \text{for} \quad 0 < y < 1$$

we can regard C in (2.3.12) as an arbitrary constant and thus (2.3.12) is the required solution satisfying the end conditions

$$\varphi(x) = \begin{cases} 0\left(|x|^{-1/2}\right) & \text{as } x \to 0 \\ 0\left(|1-x|^{-1/2}\right) & \text{as } x \to 1. \end{cases}$$

To derive the solution for the case when

$$\varphi(x) = \begin{cases} 0\left(|x|^{1/2}\right) & \text{as } x \to 0 \\ 0\left(|1-x|^{1/2}\right) & \text{as } x \to 1, \end{cases}$$

we write $\varphi(y)$ from (2.3.11) as

$$\varphi(y) = -\frac{1}{\pi} \{y(1-y)\}^{1/2} \int_0^1 \frac{f(x)}{\{x(1-x)\}^{1/2}} \frac{dx}{y-x}$$

$$+ \frac{1}{\pi} \left(\frac{y}{1-y}\right)^{1/2} \int_0^1 \frac{f(x)}{\{x(1-x)\}^{1/2}} dt. \tag{2.3.13}$$

Thus $\varphi(y)$ has the required behaviour iff the second term in (2.3.13) vanishes, i.e., $f(x)$ satisfies

$$\int_0^1 \frac{f(x)}{\{x(1-x)\}^{1/2}} dx = 0 \tag{2.3.14}$$

and in this case the solution is given by

$$\varphi(y) = -\frac{1}{\pi} \{y(1-y)\}^{1/2} \int_0^1 \frac{f(x)}{\{x(1-x)\}^{1/2}} dx. \tag{2.3.15}$$

Thus the solution of the integral equation (2.3.1) can be obtained. The results are follows:

Case (i) If $\varphi(x) = 0\left(|x-a|^{-1/2}\right)$ as $x \to a$ and $\varphi(x) = 0\left(|b-x|^{-1/2}\right)$ as $x \to b$,

then the solution of (2.3.1) is

$$\varphi(x) = -\frac{1}{\pi} \frac{1}{\{(x-a)(b-x)\}^{1/2}} \left[C + \int_a^b \{(t-a)(b-t)\}^{1/2} \frac{f(t)}{x-t} \, dt \right], \; a < x < b \quad (2.3.16)$$

where C is an arbitrary constant.

Case (ii) (a) *If* $\varphi(x) = 0\left(|x-a|^{1/2}\right)$ as $x \to a$ and $\varphi(x) = 0\left(|b-x|^{-1/2}\right)$ as $x \to b$,

then

$$\varphi(x) = -\frac{1}{\pi} \left(\frac{x-a}{b-x}\right)^{1/2} \int_a^b \left(\frac{b-t}{t-a}\right)^{1/2} \frac{f(t)}{x-t} \, dt, \; a < x < b. \quad (2.3.17)$$

(b) If $\varphi(x) = 0\left(|x-a|^{-1/2}\right)$ as $x \to a$ and $\varphi(x) = 0\left(|x-b|^{1/2}\right)$ as $x \to b$,

then

$$\varphi(x) = -\frac{1}{\pi} \left(\frac{b-x}{x-a}\right)^{1/2} \int_a^b \left(\frac{t-a}{b-t}\right)^{1/2} \frac{f(t)}{x-t} \, dt, \; a < x < b. \quad (2.3.18)$$

Case (iii) If $\varphi(x) = 0\left(|x-a|^{1/2}\right)$ as $x \to a$ and $\varphi(x) = 0\left(|x-b|^{1/2}\right)$ as $x \to b$,

Then the solution exists *if and only if* $f(x)$ satisfies

$$\int_a^b \frac{f(t)}{\{(t-a)(b-t)\}^{1/2}} \, dt = 0 \quad\quad (2.3.19)$$

(known as the *solvability criterion*) and the solution is then given by

$$\varphi(x) = -\frac{1}{\pi} \{(x-a)(b-x)\}^{1/2} \int_a^b \frac{f(t)}{\{(t-a)(b-t)\}^{1/2}} \frac{dt}{x-t}, \; a < x < b. \quad (2.3.20)$$

Note: 1 The integrals appearing in (2.3.16) to (2.3.18) and (2.3.20) are in the sense of CPV. This method was employed by Mandal and Goswami (1983), and is also given in the book by Estrada and Kanwal (2000).

2. The method of solution presented above is obviously not rigorous. There exists rigorous method of solution based on complex variable theory and can be found in the books by Muskhelishvili (1953) and Gakhov (1966). This will also be discussed in Chapter 3.

(b) *Second kind integral equations with Cauchy kernel*

We consider the simple Cauchy type singular integral equation, as given by

$$\rho \; \varphi(x) = \int_0^1 \frac{\varphi(t)}{t-x} \, dt + f(x), \; 0 < x < 1 \qquad (2.3.21)$$

where, the integral is in the sense of CPV, and for simplicity, we assume that ρ is a known constant, and $\varphi(x)$ and $f(x)$ are complex valued functions of the variable $x \in (0,1)$, $\varphi(x)$ being the unknown function of the integral equation and $f(x)$ being a known function.

It may be noted that the case $\rho = 0$, of the integral equation (2.3.21), corresponds to the integral equation of the *first kind* already considered above, whilst if $\rho \neq 0$, then the equation (2.3.21) represents a special singular integral equation of the *second kind*, with constant coefficient. There exist various complex variable methods of solutions of the integral equations of the form (2.3.21) in the literature (see Muskhelishvili (1953) and Gakhov (1966)), some aspects of which will be taken up in Chapter 3 of the book.

Here we employ the following quick and elementary method of solution of the integral equation (2.3.21), which depends on the solution of Abel type singular integral equations, described in the previous section 2.1.

We start with the following standard result

$$\int_0^1 \frac{1}{t^{1-\alpha}(1-t)^{\alpha}} \frac{dt}{t-x} = -\frac{\pi \, \cot \pi \alpha}{x^{1-\alpha}(1-x)^{\alpha}}, \; 0 < x < 1, \quad (2.3.22)$$

where α is a fixed constant such that $0 < \alpha < 1$, and the singular integral is understood in the sense of CPV.

The result (2.3.22) clearly shows that there exists a class of differentiable functions, in the open interval (0,1), which represents the solutions of the homogeneous part of the integral equation (2.3.21), in the circumstances when $\rho = -\pi \cot \pi \alpha$, and this is provided by the functions

$$\varphi_0(x) = \frac{C_0}{x^{1-\alpha}(1-x)^{\alpha}}, \; 0 < x < 1, \qquad (2.3.23)$$

where C_0 is an arbitrary constant. It may be noted that $\alpha = \dfrac{1}{2}$ when $\rho = 0$.

Guided by this observation on the homogeneous part of the integral equation (2.3.21), we now *expect* that the *general solution* of the integral equation (2.3.21) *can* be expressed in the form, as given by

$$\varphi(x) = \frac{1}{(1-x)^\alpha} \frac{d}{dx} \int_0^x \frac{\psi(t)}{(x-t)^{1-\alpha}} \, dt \qquad (2.3.24)$$

with α satisfying the relation $\rho = -\pi \cot \pi \alpha$, where $\psi(t)$ is a differentiable function with $\psi(0) \neq 0$. From (2.3.24), we find an alternative form for $\varphi(x)$ as

$$\varphi(x) = \frac{\psi(0)}{x^{1-a}(1-x)^a} + \frac{1}{(1-x)^a} \int_0^x \frac{\psi'(t)}{(x-t)^{1-a}} \, dt \qquad (2.3.25)$$

where $\psi' = \dfrac{d\psi}{dt}$. The form (2.3.25) clearly shows that the function $\varphi(x)$ possesses the same weak singularities at the end points of the interval $(0,1)$, under consideration, as is possessed by such solutions (cf.(2.3.23)) of the corresponding homogeneous equation (2.3.21).

Using the form (2.3.25) of $\varphi(x)$ we first find that

$$\int_0^1 \frac{\varphi(t)}{t-x} \, dt = \psi(0) \int_0^1 \frac{t}{t^{1-\alpha}(1-t)^\alpha} \frac{dt}{t-x} + \int_0^1 \left\{ \frac{1}{(1-t)^\alpha} \int_0^t \frac{\psi'(u)}{(t-u)^{1-\alpha}} \, du \right\} \frac{dt}{t-x}$$

$$= \psi(0) \int_0^1 \frac{1}{t^{1-\alpha}(1-t)^\alpha} \frac{dt}{t-x} + \int_0^x \psi'(u) \left\{ \int_u^1 \frac{1}{(1-t)^\alpha (t-u)^{1-\alpha}} \frac{dt}{t-x} \right\} du \qquad (2.3.26)$$

$$+ \int_x^1 \psi'(u) \left\{ \int_u^1 \frac{1}{(1-t)^\alpha (t-u)^{1-\alpha}} \frac{dt}{t-x} \right\} du$$

obtained by interchanging the orders of integration in the second term, after splitting it into two terms like $\int_0^x \cdots dt + \int_x^1 \cdots dt$. By using the following standard integrals (cf. Gakhov (1966))

$$\int_0^1 \frac{1}{(1-t)^\alpha (t-u)^{1-\alpha}} \frac{dt}{t-x} = \begin{cases} \dfrac{\pi \cos ec \, \pi \alpha}{(1-x)^\alpha (u-x)^{1-\alpha}} & \text{for } 0 < x < u < 1, \\[4mm] -\dfrac{\pi \cot \pi \alpha}{(1-x)^\alpha (x-u)^{1-\alpha}} & \text{for } 0 < u < x < 1 \end{cases} \qquad (2.3.27)$$

in the relation (2.3.26), we can express the integral equation (2.3.21) as

$$\frac{\rho\psi(0)}{x^{1-\alpha}(1-x)^{\alpha}} + \frac{\rho}{(1-x)^{\alpha}}\int_{0}^{x}\frac{\psi'(t)}{(x-t)^{1-\alpha}}dt = -\frac{\pi\,\psi(0)\cot\pi\alpha}{x^{1-\alpha}(1-x)^{\alpha}}$$

$$-\frac{\pi\cot\pi\alpha}{(1-x)^{\alpha}}\int_{0}^{x}\frac{\psi'(t)}{(x-t)^{1-\alpha}}dt + \frac{\pi\cos ec\,\pi\alpha}{(1-x)^{\alpha}}\int_{x}^{1}\frac{\psi'(t)}{(t-x)^{1-\alpha}}dt + f(x),$$

and this, on using the relation $\rho = -\pi\cot\pi\alpha$, gives rise to the following Abel type integral equation

$$\int_{x}^{1}\frac{\psi'(t)}{(t-x)^{1-\alpha}}dt = -\frac{\sin\pi\alpha}{\pi}(1-x)^{\alpha}f(x),\quad 0<x<1.\quad(2.3.28)$$

The solution of the equation (2.3.28) can be determined easily by employing the techniques described in section 2.1, and, we find that

$$\psi(x) = \frac{\sin^2\pi\alpha}{\pi^2}\int_{x}^{1}\frac{(1-y)^{\alpha}f(y)}{(y-x)^{\alpha}}dy + C\quad(2.3.29)$$

where C is an arbitrary constant of integration.

Using the representation (2.3.29) into the right side of (2.3.24), we thus determine the *general solution* of the singular integral equation (2.3.21), in the class of functions described earlier (cf.(2.3.23)), as given by

$$\varphi(x) = \frac{C}{x^{1-\alpha}(1-x)^{1-\alpha}} + \frac{\sin^2\pi\alpha}{\pi^2(1-x)^{\alpha}}\frac{d}{dx}\left[\int_{0}^{x}\frac{1}{(x-t)^{1-\alpha}}\left\{\int_{t}^{1}\frac{(1-y)^{\alpha}f(y)}{(y-t)^{\alpha}}dy\right\}dt\right].\quad(2.3.30)$$

By interchanging the order of integration on the right side of (2.3.30), and using the following results

(i) $$\int_{0}^{\min(x,y)}\frac{dt}{(x-t)^{1-\alpha}(y-t)^{\alpha}} = \begin{cases}\displaystyle\sum_{j=0}^{\infty}\frac{1}{j+\mu}\left(\frac{y}{x}\right)^{j+\mu} \equiv G_1(x,y)\text{ for }x>y\;(\mu=1-\alpha)\\[2ex]\displaystyle\sum_{j=0}^{\infty}\frac{1}{j+\alpha}\left(\frac{x}{y}\right)^{\alpha+j} \equiv G_2(x,y)\text{ for }x<y,\end{cases}$$

(ii) $$\frac{\partial G_1}{\partial x} = \left(\frac{y}{x}\right)^{\mu}\frac{1}{y-x} = \frac{\partial G_2}{\partial x},$$

(iii) $$\sin^2\pi\alpha = \frac{\pi^2}{\rho^2+\pi^2}\quad(\text{since }\rho=-\pi\cot\pi\alpha).$$

We can easily rewrite the general solution (2.3.30) of the integral equation (2.3.21), in the following well-known form (cf. Gakhov (1966))

$$\varphi(x) = \frac{\rho}{\rho^2 + \pi^2} f(x) + \frac{1}{\rho^2 + \pi^2} \frac{1}{x^{1-\alpha}(1-x)^\alpha} \int_0^1 \frac{y^{1-\alpha}(1-y)^\alpha}{y-x} f(y)\, dy + \frac{C}{x^{1-\alpha}(1-x)^\alpha}. \quad (2.3.31)$$

We observe that for the particular case $\rho = 0$, corresponding to the equation of the *first kind*, whose solution is obtained in section 2.3, the general solution can also be obtained from the relation (2.3.31), and is given by (since $\alpha = 1/2$)

$$\varphi(x) = \frac{1}{x^{1/2}(1-x)^{1/2}} \left[C + \frac{1}{\pi^2} \int_0^1 \frac{y^{1/2}(1-y)^{1/2}}{y-x} f(y)\, dy \right] \quad (2.3.32)$$

where C is an arbitrary constant. This coincides with (2.3.12) if $\frac{1}{\pi} f(y)$ is replaced by $f(y)$.

(c) *A related singular integral equation*

A related singular integral equation of the first kind with Cauchy type kernel, as given by

$$\int_0^1 \frac{\varphi(t)}{t^2 - x^2}\, dt = f(x), \quad 0 < x < 1, \quad (2.3.33)$$

which occurs in the study of the problem of surface water waves by a vertical barrier (see Mandal and Chakrabarti (1999)), can also be handled by employing the technique developed above.

We find that the substitution

$$\varphi(x) = \frac{1}{x} \frac{d}{dx} \int_x^1 \frac{t\, s(t)}{(t^2 - x^2)^{1/2}}\, dt, \quad 0 < x < 1 \quad (2.3.34)$$

where $s(t)$ is a differentiable function, with $s(1) \neq 0$, helps in solving the integral equation (2.3.33) with a differentiable forcing term $f(x)$, in the form

$$\varphi(x) = \frac{D}{(1-x^2)^{1/2}} - \frac{4}{\pi^2} \frac{1}{(1-x^2)^{1/2}} \int_0^1 \frac{t^2(1-t^2)^{1/2}}{t^2 - x^2} f(t)\, dt \quad (2.3.35)$$

where D is an arbitrary constant.

(d) *A special singular integral equation*

Next we describe a quick and elementary method of solution of the following special singular integral equation, as given by

$$a(x)(T\varphi)(x) + \big(T(b\varphi)\big)(x) = f(x), \quad -1 < x < 1 \qquad (2.3.36)$$

where the *singular integral operator* T is defined as

$$(T\varphi)(x) = \frac{1}{\pi} \int_{-1}^{1} \frac{\varphi(t)}{t-x}\, dt, \quad -1 < x < 1 \qquad (2.3.37)$$

with $a(x)$, $b(x)$ and $f(x)$ being known differentiable functions of $x \in (-1,1)$, under the circumstances when

$$a(x)b(x) = \lambda^2 (1 - x^2), \qquad (2.3.38)$$

λ being a known constant. The above *special* singular integral equation arises in the study of problems in the theory of dislocations as well as in the theory of waveguides (cf. Williams (1975), Chakrabarti and Williams (1980) and Lewin (1975)). We present below the method of Chakrabarti and Williams (1980) to determine the general solution of the integral equation (2.3.36).

We first observe that the general solution of the integral equation

$$T\varphi \equiv (T\varphi)(x) = g(x), \quad -1 < x < 1 \qquad (2.3.39)$$

can be expressed in the form (cf. equation (2.3.32))

$$\varphi(x) = (T^{-1}g)(x) \equiv \frac{C}{\left(1-x^2\right)^{1/2}} - \frac{1}{\left(1-x^2\right)^{1/2}}\, T\big((1-x^2)^{1/2} g\big)(x) \qquad (2.3.40)$$

for differentiable function g, where C is an arbitrary constant.

We notice that the operators T and T^{-1}, as defined by the relations (2.3.37) and (2.3.40) respectively, have the following three important properties:

(i)
$$T(T^{-1}\varphi) = \varphi,$$

(ii)
$$T^{-1}(T\varphi) = \frac{C}{\left(1-x^2\right)^{1/2}} + \varphi$$

and

(iii)
$$T\left(\frac{1}{\left(1-x^2\right)^{1/2}}\right) = 0.\qquad(2.3.41)$$

Thus, we have that

$$T^{-1}\left(\frac{b}{(1-x^2)^{1/2}}\varphi\right) = \frac{C}{\left(1-x^2\right)^{1/2}} - \frac{1}{\left(1-x^2\right)^{1/2}}\,T(b\varphi),\quad(2.3.42)$$

and, then, the given integral equation (2.3.36) can be cast as

$$\frac{a}{\left(1-x^2\right)^{1/2}}T\varphi - T^{-1}\left(\frac{b}{\left(1-x^2\right)^{1/2}}\varphi\right) = \frac{f}{\left(1-x^2\right)^{1/2}} - \frac{C}{\left(1-x^2\right)^{1/2}}.\quad(2.3.43)$$

Applying the operator T on both sides of the above equation (2.3.43) and using the results (2.3.41), we obtain

$$T\left(\frac{a}{\left(1-x^2\right)^{1/2}}T\varphi\right) - \frac{b}{\left(1-x^2\right)^{1/2}}\varphi = T\left(\frac{f}{\left(1-x^2\right)^{1/2}}\right)\quad(2.3.44)$$

from which it follows that

$$\frac{a}{\left(1-x^2\right)^{1/2}}T\left(\frac{a}{\left(1-x^2\right)^{1/2}}T\varphi\right) - \frac{ab}{1-x^2}\varphi = \frac{a}{\left(1-x^2\right)^{1/2}}T\left(\frac{f}{\left(1-x^2\right)^{1/2}}\right).\quad(2.3.45)$$

If we now define a new operator L, as given by

$$L \equiv \frac{a(x)}{\left(1-x^2\right)^{1/2}}T,\qquad(2.3.46)$$

and utilize the relation (2.3.39), we find that the equation (2.3.45) can be expressed as

$$\left(L^2 - \lambda^2\right)\varphi = h \qquad (2.3.47)$$

where

$$h(x) = \frac{a(x)}{\left(1 - x^2\right)^{1/2}} \, T\left(\frac{f}{\left(1 - x^2\right)^{1/2}}\right). \qquad (2.3.48)$$

The equation (2.3.47) can now be cast into either of the following two forms:

$$\left((L - \lambda)\psi_1\right)(x) = h(x), \qquad (2.3.49)$$

or,

$$\left((L + \lambda)\psi_2\right)(x) = h(x) \qquad (2.3.50)$$

where

$$\psi_1(x) = \left((L + \lambda)\varphi\right)(x) \qquad (2.3.51)$$

and

$$\psi_2(x) = \left((L - \lambda)\varphi\right)(x) \qquad (2.3.52)$$

so that the unknown function $\varphi(x)$, of our concern, can be expressed as

$$\varphi(x) = \frac{1}{2\lambda} \left(\psi_1(x) - \psi_2(x)\right). \qquad (2.3.53)$$

Utilizing the operator L, as defined by the relation (2.3.46), the integral equation (2.3.49) can be expressed as

$$T\psi_1 - \lambda \frac{\left(1 - x^2\right)^{1/2}}{a}\psi_1 = T\left(\frac{f}{\left(1 - x^2\right)^{1/2}}\right)$$

which, on applying the operator T^{-1} to both sides, produces (see equations (2.3.41))

$$\psi_1 + \frac{\lambda}{\left(1 - x^2\right)^{1/2}} \, T\left(\frac{1 - x^2}{a}\psi_1\right) = \frac{f}{\left(1 - x^2\right)^{1/2}} + \frac{A}{\left(1 - x^2\right)^{1/2}} \qquad (2.3.54)$$

where A is an arbitrary constant.

Now, if we define

$$\mu(x) = \frac{\left(1-x^2\right)^{1/2}}{a(x)} \quad \text{and} \quad \Psi_1(x) = \left(1-x^2\right)\psi_1(x), \quad (2.3.55)$$

then the equation (2.3.54) takes up the form

$$\Psi_1(x) + \lambda T(\mu \Psi_1)(x) = f(x) + A, \quad -1 < x < 1. \quad (2.3.56)$$

A *similar* analysis, applied to the equation (2.3.50), produces the equation

$$\Psi_2(x) - \lambda T(\mu \Psi_2)(x) = f(x) + B, \quad -1 < x < 1 \quad (2.3.57)$$

where

$$\Psi_2(x) = \left(1-x^2\right)^{1/2} \psi_2(x) \quad (2.3.58)$$

and B is an arbitrary constant, different from A.

We thus observe that the original singular integral equation (2.3.36) can be solved by way of solving two independent singular integral equations (2.3.56) and (2.3.57), and utilizing the algebraic relation (2.3.53). We consider below a special case of the above general problem of singular integral equation.

A special case

In the special case when

$$a(x) = \lambda^2 \left(1-x^2\right)^{1/2} \quad \text{and} \quad b(x) = \left(1-x^2\right)^{1/2}, \quad (2.3.59)$$

the function $\mu(x)$ in the relation (2.3.55) becomes

$$\mu(x) = \frac{1}{\lambda^2} \quad (2.3.60)$$

which is a *constant*, and then the two integral equations (2.3.56) and (2.3.57) take up the following simple forms:

$$\Psi_1(x) + \frac{1}{\lambda}(T\Psi_1) = f(x) + A, \quad -1 < x < 1 \quad (2.3.61)$$

and

$$\Psi_2(x) - \frac{1}{\lambda}(T\Psi_2) = f(x) + B, \quad -1 < x < 1. \quad (2.3.62)$$

The solutions of the above two singular integral equations (2.3.61) and (2.3.62) can be easily obtained, in the case when $\lambda > 0$, by using a variation of the result (2.3.31), and we find that

$$\Psi_1(x) = \frac{\lambda^2}{1+\lambda^2}(f(x)+A) - \frac{\lambda}{1+\lambda^2}\left(\frac{1+x}{1-x}\right)^{\frac{1}{2}-\beta}\frac{1}{\pi}\int_{-1}^{1}\left(\frac{1-t}{1+t}\right)^{\frac{1}{2}-\beta}\frac{f(t)+A}{t-x}\,dt$$
$$+ \frac{K_1}{(1+x)^{\frac{1}{2}+\beta}(1-x)^{\frac{1}{2}-\beta}}. \tag{2.3.63}$$

and

$$\Psi_2(x) = \frac{\lambda^2}{1+\lambda^2}(f(x)+B) + \frac{\lambda}{1+\lambda^2}\left(\frac{1-x}{1+x}\right)^{\frac{1}{2}-\beta}\frac{1}{\pi}\int_{-1}^{1}\left(\frac{1+t}{1-t}\right)^{\frac{1}{2}-\beta}\frac{f(t)+B}{t-x}\,dt$$
$$+ \frac{K_2}{(1+x)^{\frac{1}{2}-\beta}(1-x)^{\frac{1}{2}+\beta}}, \tag{2.3.64}$$

where

$$\lambda = \tan\pi\beta \quad (\lambda > 0) \tag{2.3.65}$$

and, K_1 and K_2 are arbitrary constants.

We note that in the most special case of the above problem, where the constants A, B, K_1 and K_2 are all zeros, the solution of the singular integral equation (2.3.36) can be finally expressed as

$$\varphi(x) = -\frac{1}{2(1+\lambda^2)}\frac{1}{1-x}\left(\frac{1+x}{1-x}\right)^{-\beta}T\left(\left(\frac{1+x}{1-x}\right)^{\beta-\frac{1}{2}}f\right)$$
$$+ \left(\frac{1+x}{1-x}\right)^{\beta-1}T\left(\left(\frac{1+x}{1-x}\right)^{\frac{1}{2}-\beta}f\right). \tag{2.3.66}$$

Remark

In solving the singular integral equations (2.3.2) and (2.2.21) with Cauchy kernels, solutions of Abel integral equations have been utilized. This idea of use of Abel integral equations in solving Cauchy singular integral equations of first kind was in fact originally described by Peters (1963). The function $\varphi(x)$ is assumed to satisfy a uniform Hölder condition in the closed interval $[0,1]$ and at the end points $x = 0,1$, it may have a singularity like $\ln|\alpha - x|$ or $(\alpha - x)^{-\gamma}$ $(0 < \gamma < 1)$ where α denotes any one of the end points. The forcing function $f(x)$ is also assumed to be a member of the class of functions to which $\varphi(x)$ belongs.

Peters (1963) used the following important idea, which is useful for any integral equation of the first kind, as given by

$$\int_0^1 K(x,t)\, \varphi(t)\, dt = f(x),\ 0 < x < 1 \tag{2.3.67}$$

for which the kernel $K(x,t)$ possesses the representation

$$K(x,t) = \begin{cases} \displaystyle\int_0^t K_1(x,\sigma)\, K_2(t,\sigma)\, d\sigma,\ t < x, \\[4mm] \displaystyle\int_0^x K_1(x,\sigma)\, K_2(t,\sigma)\, d\sigma,\ x < t. \end{cases} \tag{2.3.68}$$

Then, using the representation (2.3.68), the integral equation (2.3.67) can be reduced to

$$\int_0^x K_1(x,\sigma)\left(\int_\sigma^1 K_2(t,\sigma)\, \varphi(t)\, dt \right) d\sigma = f(x),\ 0 < x < 1 \tag{2.3.69}$$

which essentially represents two independent integral equations of Volterra type, as given by

$$\int_0^x K_1(x,\sigma)\, \psi(\sigma)\, d\sigma = f(x),\ 0 < x < 1 \tag{2.3.70}$$

and

$$\int_\sigma^1 K_2(t,\sigma)\, \varphi(t)\, dt = \psi(\sigma),\ 0 < \sigma < 1. \tag{2.3.71}$$

We note that the most important relation $\psi(1) = 0$ must be satisfied. Thus, the original problem of determining the unknown function $\varphi(x)$ satisfying the integral equation (2.3.67) can be successfully resolved by solving the two integral equations (2.3.70) and (2.3.71) on section 2.3(a) to solve the Cauchy singular integral equation (2.3.1).

2.4 APPLICATION TO BOUNDARY VALUE PROBLEMS IN ELASTICITY AND FLUID MECHANICS

As applications of the solutions of singular integral equations to boundary value problems in elasticity and fluid mechanics, we take up in this section two problems formulated in Chapter 1.

(a) *A crack problem in the theory of elasticity*

It has already been shown in section 1.2.2 that the problem of determination of the distribution of stress in the vicinity of a Griffith crack $|x| < 1$, $y = 0$ (in Cartesian xy-co-ordinates) in an infinite isotropic elastic plate, can be solved completely, by using the solution of the singular integral equation of the Cauchy type given by (1.2.23) which is required to be solved under the end conditions given by (1.2.24). Using the form (2.3.18) (with $a = 0$, $b = 1$ and $f(x)$ replaced by $\dfrac{f(x)}{\pi}$) we find that the appropriate solution of the integral equation (1.2.23) satisfying the requirements (1.2.24), is given by

$$f(s) = -\frac{1}{\pi^2}\left(\frac{1-s}{s}\right)^{1/2} \int_0^1 \left(\frac{t}{1-t}\right)^{1/2} \frac{g(t)}{t-s}\, dt. \qquad (2.4.1)$$

It is a matter of some routine manipulations to determine the distribution of stress $\sigma_{xx}, \sigma_{xy}, \sigma_{yy}$ as given by the relations (1.2.12), for the special boundary value problem of elasticity considered here, when use is made of the relations (1.2.21), (1.2.18) along with the knowledge of the functions $f(s)$, as given by (2.4.1).

(b) *A surface wave scattering problem*

As has been shown in section 1.2.3, the problem of determination of the scattered potential, in the linearised theory of water waves, when a train of surface water waves is normal incident on a thin vertical barrier partially immersed in infinitely deep water, can be solved completely by reducing it to the homogeneous singular integral equation (1.2.61) under the end conditions (1.2.62). Its appropriate solution is now obtained by using the solution (2.3.16) (with $a = 0$, $b = 1$, $f(x) = 0$) and is given by

$$q(u) = \frac{C}{u^{1/2}(1-u)^{1/2}} \qquad (2.4.2)$$

where C is an arbitrary constant. Thus the function $p(t)$ defined by (1.2.60) is found to be

$$p(t) = \frac{Ct}{\left(t^2 - a^2\right)^{1/2}} \tag{2.4.3}$$

and thus the function $g(y)$ satisfying the integral equation (1.2.52) is obtained as

$$g(y) = C\frac{d}{dy}\left\{ e^{-Ky} \int_a^y \frac{u\ e^{Ku}}{\left(u^2 - a^2\right)^{1/2}} du \right\}, \quad u > a. \tag{2.4.4}$$

The unknown constant C, the complex reflection and transmission coefficients R and T, and the scattered potential can then be determined by utilizing the various connecting relations given in section 1.2.3, which however is not presented here. Details can be found in the book of Mandal and Chakrabarti (2000).

Chapter 3

Riemann-Hilbert Problems and Their Uses in Singular Integral Equations

In this chapter we describe the analysis as well as the methods of solution of a special type of problems of complex variable theory, called Riemann-Hilbert problems (RHP). It will be shown here that converting the equations to RHPs and finally solving them can, successfully solve the singular integral equations involving the Cauchy type singularities in their kernels. This method has already been introduced in Chapter 2 in an *ad hoc* manner to solve a singular integral equation of some special type involving logarithmic type kernel. This method is also known in the literature as *function-theoretic method*.

Examples of singular integral equations occurring in Elasticity, Fluid Mechanics and related areas, will be considered and the detailed analysis to solve some of the singular integral equations arising in these areas will be explained. The Cauchy type singular integral equations have already been introduced in chapter 1 briefly and solutions of some of them have been obtained by some elementary methods. We have tried to present all the basic ideas needed to implement the analysis involving RHPs in as simple a manner as has been possible, so that even beginners can understand easily.

3.1 CAUCHY PRINCIPAL VALUE INTEGRALS

In this section we explain the ideas involving a special class of integrals, which are singular and are of the Cauchy type. These have already been introduced in section 1.1 while defining integral equation with Cauchy type singular kernel. As an example of a Cauchy type singular integral, we consider the integral

$$I = \fint_a^b \frac{1}{x-c} \, dx, \ a < c < b \ \text{ with } \ a, b \in \mathbb{R}. \qquad (3.1.1)$$

The integral (3.1.1) does not exist in the usual sense, but if we interpret I as

$$I = \lim_{\varepsilon \to +0} \left[\int_{a}^{c-\varepsilon} \frac{1}{x-c} \, dx + \int_{c+\varepsilon}^{b} \frac{1}{x-c} \, dx \right], \qquad (3.1.2)$$

then we find that

$$I = \ln \left| \frac{b-c}{c-a} \right| \qquad (3.1.3)$$

which is an well defined quantity. This is taken to be the Cauchy principal value of the integral under consideration, to be denoted with a *cut* across the sign of integration, as in the relation (3.1.1). All such singular integrals appearing in this book will be understood to have similar meaning and we simply write

$$I = \int_{a}^{b} \frac{1}{x-c} \, dx,$$

where the *cut* across the sign of integration is withdrawn.

The most general singular integral of the Cauchy type is the one given by the relation

$$\hat{I} = \int_{\Gamma} \frac{f(z)}{z-\zeta} \, dz = \lim_{\varepsilon \to +0} \int_{\Gamma - \Gamma_{\varepsilon}} \frac{f(z)}{z-\zeta} \, dz \qquad (3.1.4)$$

where Γ is a smooth contour in the complex z-plane and ζ is a point on the contour Γ (see Figure 3.1.1)

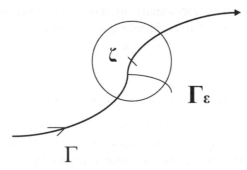

Fig. 3.1.1 Contour Γ

and Γ_{ε} is the portion of the contour Γ which lies inside a circle of radius ε centred at $z = \zeta$.

We emphasize here that many boundary value problems of mathematical physics can be transformed into singular integral equations where the kernels are of Cauchy type (cf. Sneddon (1974), Gakhov (1966), Muskhelishvilli (1953), Ursell (1947), Mandal and Chakrabarti (2000) and others).

3.2 SOME BASIC RESULTS IN COMPLEX VARIABLE THEORY

We now state some important theorems and results in the theory of functions of complex variables, without their detailed proofs, which can be found in Muskhelishvilli (1953), Gakhov (1966).

Theorem 3.2.1

For the integral

$$\Phi(z) = \frac{1}{2\pi i} \int_{\Gamma} \frac{\varphi(\tau)}{\tau - z} d\tau, \quad z \notin \Gamma, \tag{3.2.1}$$

if the density function $\varphi(\tau)$ satisfies the Hölder condition

$$|\varphi(\tau_1) - \varphi(\tau_2)| < A|\tau_1 - \tau_2|^{\alpha}, \quad 0 < \alpha < 1 \tag{3.2.2}$$

with A being a positive constant, for all pairs of points τ_1, τ_2 on a *simple closed positively oriented contour* Γ of the complex z-plane ($z = x + iy$), then $\Phi(z)$ represents a sectionally analytic (analytic except for points z lying on Γ) function of the complex variable z.

Theorem 3.2.2 (The Basic Lemma)

The function

$$\psi(z) = \frac{1}{2\pi i} \int_{\Gamma} \frac{\varphi(\tau) - \varphi(t)}{\tau - z} d\tau \tag{3.2.3}$$

on passing through the point $z = t$, of the simple closed contour Γ, behaves as a continuous function of z, i.e.

$$\lim_{z \to t} \psi(z) = \frac{1}{2\pi i} \int_{\Gamma} \frac{\varphi(\tau) - \varphi(t)}{\tau - t} d\tau \tag{3.2.4}$$

exists and is equal to $\psi(t)$, whenever φ satisfies a Hölder condition on Γ.

Note: The theorem 3.2.2 also holds at every point on Γ, except at the end points, when Γ is an open arc in the complex z-plane.

Theorem 3.2.3 (Plemelj-Sokhotski formulae)

If

$$\Phi(z) = \frac{1}{2\pi i} \int_{\Gamma} \frac{\varphi(\tau)}{\tau - z}\, d\tau,\ \ z \notin \Gamma,$$

with φ satisfying a Hölder condition on Γ, then

$$\lim_{z \to t+} \Phi(z) = \Phi^+(t)\ \text{ and }\ \lim_{z \to t-} \Phi(z) = \Phi^-(t)$$

exist, and the following formulae hold good:

$$\Phi^+(t) - \Phi^-(t) = \varphi(t),\ \ t \in \Gamma, \tag{3.2.5a}$$

$$\Phi^+(t) + \Phi^-(t) = \frac{1}{\pi i} \int_{\Gamma} \frac{\varphi(\tau)}{\tau - t}\, d\tau,\ \ t \in \Gamma \tag{3.2.5b}$$

where $\lim\limits_{z \to t+}$ and $\lim\limits_{z \to t-}$ mean that the point z approaches the point t on Γ from the left side and from the right side respectively of the positively oriented contour Γ, with the singular integral appearing above being in the sense of CPV.

The formulae (3.2.5) are known as the *Plemelj formulae* (also referred to as the *Sokhotski formulae*) involving the Cauchy type integral $\Phi(z)$, which can also be expressed as

$$\Phi^{\pm}(t) = \pm\, \frac{1}{2}\varphi(t) + \frac{1}{2\pi i} \int_{\Gamma} \frac{\varphi(\tau)}{\tau - t}\, d\tau,\ \ t \in \Gamma. \tag{3.2.6}$$

Note: The Plemelj formulae also hold good even if Γ is an arc (or a finite union of arcs) provided that t does not coincide with an end point of Γ.

Proof: We can easily prove the Plemelj formulae in the case when Γ is a *closed* smooth contour, by using the following results:

$$\frac{1}{2\pi i} \int_{\Gamma} \frac{1}{\tau - z} \, d\tau = \begin{cases} 1, & z \in D^+ \\ 0, & z \in D^- \\ \dfrac{1}{2}, & z \in \Gamma, \end{cases} \tag{3.2.7}$$

where D^+ is the region lying inside the simple closed contour Γ and D^- is the region lying outside F (see Figure 3.2.1).

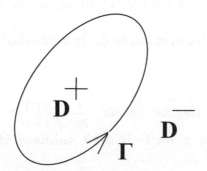

Fig. 3.2.1 Regions D^+ and D^-

Writing

$$\psi(z) = \frac{1}{2\pi i} \int_{\Gamma} \frac{\varphi(\tau) - \varphi(t)}{\tau - z} \, d\tau$$

we find that

$$\lim_{z \to t+} \psi(z) = \lim_{z \to t+} \frac{1}{2\pi i} \int_{\Gamma} \frac{\varphi(\tau)}{\tau - z} \, d\tau - \varphi(t) \lim_{z \to t+} \frac{1}{2\pi i} \int_{\Gamma} \frac{1}{\tau - z} \, d\tau$$

giving, on use of the results (3.2.7),

$$\psi^+(t) = \Phi^+(t) - \varphi(t), \quad t \in \Gamma. \tag{3.2.8}$$

Also,

$$\lim_{z \to t-} \psi(z) = \lim_{z \to t-} \frac{1}{2\pi i} \int_{\Gamma} \frac{\varphi(\tau)}{\tau - z} \, d\tau - \varphi(t) \lim_{z \to t-} \frac{1}{2\pi i} \int_{\Gamma} \frac{1}{\tau - z} \, d\tau$$

giving

$$\psi^-(t) = \Phi^-(t) - 0 = \Phi^-(t), \quad t \in \Gamma. \tag{3.2.9}$$

Now, we have

$$\psi(t) = \frac{1}{2\pi i} \int_\Gamma \frac{\varphi(t)}{\tau - t} \, d\tau - \varphi(t) \frac{1}{2\pi i} \int_\Gamma \frac{1}{\tau - t} \, d\tau, \quad t \in \Gamma$$

which gives

$$\psi(t) = -\frac{1}{2} \varphi(t) + \frac{1}{2\pi i} \int_\Gamma \frac{\varphi(\tau)}{\tau - t} \, d\tau, \quad t \in \Gamma, \qquad (3.2.10)$$

when the result (3.2.7) is utilized. Now the Theorem 3.2.2 suggests that all the results (3.2.8), (3.2.9) and (3.2.10) are identical and we thus derive the results (3.2.6).

The following theorem can be easily established (we omit the proof here).

Theorem 3.2.4

The Cauchy type singular integral $\dfrac{1}{2\pi i} \int_\Gamma \dfrac{\varphi(\tau)}{\tau - z} \, d\tau$ satisfies a Hölder condition for points z on Γ if $\varphi(\tau)$ satisfies a Hölder condition for points τ on Γ.

We next establish an important formula, known as *Poincare'-Bertrand Formula*, involving singular integrals, as explained below in the form of a theorem.

Theorem 3.2.5 (Poincare'-Bertrand Formula (PBF))

If Γ is a simple closed contour, and if φ satisfies a Hölder condition on Γ, then the PBF

$$\int_\Gamma \frac{1}{\tau - t} \left\{ \int_\Gamma \frac{\varphi(s)}{s - \tau} \, ds \right\} d\tau = -\pi^2 \varphi(t), \quad t \in \Gamma, \qquad (3.2.11)$$

holds good.

Proof: We set

$$\varphi_1(t) = \frac{1}{2\pi i} \int_\Gamma \frac{\varphi(\tau)}{\tau - t} \, d\tau, \quad t \in \Gamma, \qquad (3.2.12)$$

$$\varphi_2(t) = \frac{1}{2\pi i} \int_\Gamma \frac{\varphi_1(\tau)}{\tau - t} \, d\tau, \quad t \in \Gamma, \qquad (3.2.13)$$

and also

$$\Phi(z) = \frac{1}{2\pi i} \int_{\Gamma} \frac{\varphi(\tau)}{\tau - z} d\tau, \quad z \notin \Gamma, \tag{3.2.14}$$

$$\Phi_1(z) = \frac{1}{2\pi i} \int_{\Gamma} \frac{\varphi_1(\tau)}{\tau - z} d\tau, \quad z \notin \Gamma, \tag{3.2.15}$$

Then using the Plemelj formulae (3.2.6), we obtain

$$\varphi_1(t) = \Phi^+(t) - \frac{1}{2} \varphi(t), \quad t \in \Gamma, \tag{3.2.16}$$

$$\varphi_2(t) = \Phi_1^+(t) - \frac{1}{2} \varphi_1(t), \quad t \in \Gamma. \tag{3.2.17}$$

Then, using (3.2.16) in (3.2.15) we obtain

$$\Phi_1(z) = \frac{1}{2\pi i} \int_{\Gamma} \frac{\Phi^+(\tau)}{\tau - z} d\tau - \frac{1}{4\pi i} \int_{\Gamma} \frac{\varphi(\tau)}{\tau - z} d\tau$$

$$= \Phi(z) - \frac{1}{2} \Phi(z), \quad z \in D^+,$$

giving

$$\Phi_1(z) = \frac{1}{2} \Phi(z), \quad z \in D^+. \tag{3.2.18}$$

Thus we derive that

$$\Phi_1^+(t) = \frac{1}{2} \Phi^+(t), \quad t \in \Gamma. \tag{3.2.19}$$

Using (3.2.19) in the relations (3.2.16) and (3.2.17), we deduce that

$$\varphi_2(t) = \frac{1}{4} \varphi(t), \quad t \in \Gamma. \tag{3.2.20}$$

The relations (3.2.12), (3.2.13) and (3.2.20) prove the Poincaré-Bertrand Formula (3.2.11), finally.

Note: 1. The Poincaré-Bertrand Formula is useful whenever the orders of repeated singular integrals are interchanged.

2. Another form of this theorem is given later in chapter 4.

3.3 SOLUTION OF SINGULAR INTEGRAL EQUATIONS INVOLVING CLOSED CONTOURS

In this section we explain the procedure to solve a singular integral equation involving simple closed contour by means of a simple example only. Another example is given as an exercise.

Example 3.3.1

Solve the singular integral equation

$$t(t-2)\,\varphi(t) + \frac{t^2 - 6t + 8}{\pi i} \int_\Gamma \frac{\varphi(\tau)}{\tau - t}\, d\tau = \frac{1}{t}, \quad t \in \Gamma, \quad (3.3.1)$$

where Γ is a simple closed contour enclosing the origin but the point $z = 2$ lies outside Γ as shown in the Figure 3.3.1.

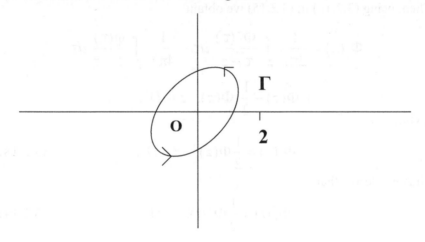

Fig. 3.3.1 The curve Γ

We set

$$\Phi(z) = \frac{1}{2\pi i} \int_\Gamma \frac{\varphi(\tau)}{\tau - z}\, d\tau, \quad z \in \Gamma \quad (3.3.2)$$

Using the Plemelj formulae (3.2.6), we can express the given equation (3.3.1) as

$$(t-2)\Phi^+(t) - 2\Phi^-(t) = \frac{1}{2t(t-2)}, \quad t \in \Gamma. \quad (3.3.3)$$

Multiplying both sides of (3.3.3) by $\dfrac{1}{2\pi i}\dfrac{1}{t-z}$ and integrating with respect to t over the contour Γ, we easily find that

$$\Phi(z) = \begin{cases} \dfrac{1}{4(z-2)^2}, & z \in D^+ \\[4mm] \dfrac{1}{8z}, & z \in D^-. \end{cases} \tag{3.3.4}$$

Hence we obtain the solution $\varphi(t)$ of the integral equation (3.3.1) as

$$\varphi(t) = \Phi^+(t) - \Phi^-(t)$$
$$= \frac{1}{4(t-2)^2} - \frac{1}{8t} = \frac{6t - t^2 - 4}{8t(t-2)^2}. \tag{3.3.5}$$

Note: The above example is taken from Chakrabarti (2008, p 130). Further examples may be found in the books of Muskhelishvilli (1953) and Gakhov (1966).

Exercise 3.3.2 Solve the singular integral equation

$$(t+1)(t-2)\varphi(t) - \frac{t^2 - 5t + 6}{\pi i} \int_{\Gamma} \frac{\varphi(\tau)}{\tau - t} d\tau = \frac{4}{t}, \quad t \in \Gamma$$

where Γ is the same as in Figure 3.3.1. This is left as an exercise.

3.4 RIEMANN-HILBERT PROBLEMS

Singular integral equations involving open arcs

The theory of a single linear singular integral equation of the second kind and of the type

$$c(t)\varphi(t) + \int_{\Gamma} \frac{\varphi(\tau)}{\tau - t} d\tau = f(t), \quad t \in \Gamma, \tag{3.4.1}$$

where $c(t)$, $f(t)$ and $\varphi(t)$ are Hölder continuous functions on Γ with Γ being a finite union of *open* arcs, can be developed as explained below (cf. Muskhelishvilli (1953), Gakhov (1966)).

We define the sectionally analytic function

$$\Phi(z) = \frac{1}{2\pi i} \int_\Gamma \frac{\varphi(\tau)}{\tau - z} \, d\tau, \ z \notin \Gamma. \tag{3.4.2}$$

Using the Plemelj formulae (3.2.6) (it can be proved that the formulae (3.2.6) are true if Γ is a finite union of open arcs), we express the integral equation (3.4.1) as a linear combination of Φ^+ and Φ^-, as given by the relation

$$c(t)\left[\Phi^+(t) - \Phi^-(t)\right] + \pi i\left[\Phi^+(t) + \Phi^-(t)\right] = f(t), \ t \in \Gamma,$$

i.e.,

$$\Phi^+(t) = \frac{c(t) - \pi i}{c(t) + \pi i} \, \Phi^-(t) + \frac{f(t)}{c(t) + \pi i}, \ t \in \Gamma, \tag{3.4.3}$$

provided $c(t) \neq -\pi i$.

We observe that the relation (3.4.3) is of the form

$$\Phi^+(t) = G(t)\Phi^-(t) + g(t), \ t \in \Gamma, \tag{3.4.4}$$

where $G(t)$ and $g(t)$ are Hölder continuous functions on Γ. The problem (3.4.4) involving the sectionally analytic function $\Phi(z)$ is called the RHP. We now describe some salient features of RHP.

Statement of Riemann-Hilbert problem

The *Riemann-Hilbert problem* is to determine a sectionally analytic function $\Phi(z)$, defined in the whole of the complex z-plane ($z = x + iy$, $x, y \in \mathbb{R}$), cut along Γ (a union of finite number of simple, smooth, non-intersecting positively oriented (anticlockwise) arcs (contours), the ends of the arcs being called *end points*), with prescribed behaviour at $z = \infty$, satisfying either of the following boundary conditions on Γ:

(i)
$$\Phi^+(t) = G(t)\Phi^-(t), \ t \in \Gamma,$$

or

(ii)
$$\Phi^+(t) = G(t)\Phi^-(t) + g(t), \ t \in \Gamma,$$

where $G(t)$ and $g(t)$ satisfy Hölder condition on Γ and $G(t) \neq 0$ for all $t \in \Gamma$.

The solutions of the problems, as posed by the equations (i) (the *homogeneous* RHP) and (ii) (the *nonhomogeneous* RHP), under sufficiently general conditions, are beyond the scope of the present book. However, under special circumstances giving rise to specially simple values of the two functions $G(t)$ and $g(t)$, the following simplified idea seems to be sufficient for the purpose of finding the solutions of the RHPs described by (i) and (ii).

Method of solution of the RHP

Let us denote by $\Phi_0(z)$ as a solution of the RHP (i) so that $\Phi_0(z)$ satisfies

$$\Phi_0^+(t) = G(t)\Phi_0^-(t), \quad t \in \Gamma.$$

Taking logarithms of both sides of the homogeneous equation, we obtain

$$ln\ \Phi_0^+(t) - ln\ \Phi_0^-(t) = ln\ G(t), \quad t \in \Gamma. \tag{3.4.5}$$

We now observe that there may exist a particular solution $\Phi_0(z)$ of the equation (3.4.5) such that

$$\left[ln\ \Phi_0\right]^+(t) - \left[ln\ \Phi_0\right]^-(t) = ln\ G(t), \quad t \in \Gamma, \tag{3.4.6}$$

giving a possible solution of (3.4.6), after noting the relation (3.2.5a), as

$$ln\ \Phi_0(z) = \frac{1}{2\pi i} \int_\Gamma \frac{ln\ G(t)}{t - z}\ dt. \tag{3.4.7}$$

It must be emphasized that the function $\Phi_0(z)$ obtained from the relation (3.4.7) is a very special solution of the *homogeneous* RHP (I), for which the two relations (3.4.5) and (3.4.6) are equivalent, and that we can always add *any entire* function of z to the expression for $ln\ \Phi_0(z)$, derived by using the relation (3.4.7), and still obtain a solution of the homogeneous RHP (i).

Let us assume that we are using the function $\Phi_0(z)$, as obtained in the relation (3.4.7). Then using the fact that

$$G(t) = \frac{\Phi_0^+(t)}{\Phi_0^-(t)}, \quad t \in \Gamma, \tag{3.4.8}$$

we can rewrite the nonhomogeneous RHP (ii) as

$$\frac{\Phi^+(t)}{\Phi_0^+(t)} - \frac{\Phi^-(t)}{\Phi_0^-(t)} = \frac{g(t)}{\Phi_0^+(t)}, \quad t \in \Gamma \tag{3.4.9}$$

using the fact that $\Phi_0^+(t) \neq 0$ for $t \in \Gamma$.

The relation (3.4.9) again represents a Riemann-Hilbert problem of a very special type, and its solution can be obtained by noting the Plemelj formula (3.2.5a), in the form given by

$$\Phi(z) = \Phi_0(z) \left[\frac{1}{2\pi i} \int_\Gamma \frac{g(t)}{\Phi_0^+(t)(t-z)} \, dt + E(z) \right] \tag{3.4.10}$$

where $E(z)$ is an entire function of z, in the whole of the complex z-plane, including Γ.

In sections 3.5, 3.6 and 3.7 we will discuss certain singular integral equations with weak singularities such as Abel and logarithmic, and show that such equations can be solved by reducing them to RHPs.

We now go back to obtaining the solution of RHP (3.4.3), which is equivalent to the singular integral equation (3.4.1). We consider, for simplicity, the case of the singular integral equation (3.4.1) in which $c(t) = \rho$, a real positive constant, and Γ is the open interval $(0,1)$ of the real axis of the complex z-plane, where $z = x + iy$. The *homogeneous* Riemann-Hilbert problem in this case, has the form

$$\Phi_0^+(x) = \frac{\rho - \pi i}{\rho + \pi i} \, \Phi_0^-(x), \quad x \in (0,1). \tag{3.4.11}$$

Now, with the aid of any *suitable* solution $\Phi_0(z)$, of the homogeneous problem (3.4.11), the original Riemann-Hilbert problem can be expressed as

$$\frac{\Phi^+(x)}{\Phi_0^+(x)} - \frac{\Phi^-(x)}{\Phi_0^-(x)} = \frac{f(x)}{(\rho + \pi i)\Phi^+(x)}, \quad x \in (0,1). \tag{3.4.12}$$

We observe that the relation (3.4.12) represents a very special Riemann-Hilbert problem for the determination of the sectionally analytic function $\Phi(z)/\Phi_0(z)$, and for a particular class of functions $f(x)$

involved in the forcing term of the original integral equation (3.4.1) under our consideration, for $x \in \Gamma = (0,1)$.

The general solution of the Riemann-Hilbert problem (3.4.12) can be written as

$$\Phi(z) = \Phi_0(z)\left[\frac{1}{2\pi i}\frac{1}{\rho + \pi i}\int_0^1 \frac{f(t)}{\Phi_0^+(t)(t-z)}\,dt + E(z)\right], \quad (3.4.13)$$

where $E(z)$ is an arbitrary entire function of z. Then, the solution of the integral equation (3.4.1) can be derived from the relation

$$\varphi(x) = \Phi^+(x) - \Phi^-(x), \quad 0 < x < 1. \qquad (3.4.14)$$

We thus find that the general solution of the integral equation (3.4.1) depends on an arbitrary entire function $E(z)$ appearing in the relation (3.4.13).

We illustrate the above procedure, for the special case of the function $\varphi(x)$ and $f(x)$, which are such that $\varphi(x)$ and $f(x)$ are bounded at $x = 0$ but have integrable singularities at $x = 1$.

Now, if we choose

$$\Phi_0(z) = \left(\frac{z}{z-1}\right)^\alpha \quad (0 < \alpha < \frac{1}{2}) \qquad (3.4.15)$$

where

$$\frac{\rho - \pi i}{\rho + \pi i} = e^{-2\pi i \alpha} \qquad (3.4.16)$$

so that

$$\rho = \pi \cot \pi\alpha, \qquad (3.4.17)$$

then, by fixing the idea that $0 < \arg z \le 2\pi$, we find

$$\Phi_0^+(x) = \left(\frac{x}{1-x}\right)^\alpha e^{i\pi\alpha}, \quad \Phi_0^-(x) = \left(\frac{x}{1-x}\right)^\alpha e^{3i\pi\alpha}. \qquad (3.4.18)$$

Also,

$$\lim_{z \to \infty} \Phi_0(z) = 1 \qquad (3.4.19)$$

and

$$\lim_{z \to \infty} \Phi(z) = 0 \qquad (3.4.20)$$

as suggested by the relation (3.4.2). Using these in (3.4.13) we find that we must have

$$E(z) \equiv 0, \tag{3.4.21}$$

giving

$$\Phi(z) = \frac{\Phi_0(z)}{2\pi i} \frac{1}{\rho + \pi i} \int_0^1 \frac{f(t)}{\Phi_0^+(t)(t-z)} \, dt, \quad z \notin (0,1). \tag{3.4.22}$$

Then utilizing the Plemelj formulae on the relation (3.4.22), together with the results (3.4.18), we find that

$$\varphi(x) = \Phi^+(x) - \Phi^-(x)$$

$$= \frac{\Phi_0^+(x) - \Phi_0^-(x)}{2\pi i(\rho + \pi i)} \int_0^1 \frac{f(t)}{\Phi_0^+(t)(t-x)} \, dt + \frac{\Phi_0^+(x) + \Phi_0^-(x)}{2\Phi_0^+(x)(\rho + \pi i)} f(x)$$

which simplifies to

$$\varphi(x) = \frac{1}{\rho^2 + \pi^2} \left[\rho f(x) - \left(\frac{x}{1-x} \right)^\alpha \int_0^1 \left(\frac{1-t}{t} \right)^\alpha \frac{f(t)}{t-x} \, dt \right], \quad 0 < x < 1. \tag{3.4.23}$$

Notes: 1. The limiting case $\rho = 0$ (i.e. $\alpha = 1/2$) of the integral equation (3.4.1) with $\Gamma = (0,1)$ is the integral equation of first kind as given by

$$\int_0^1 \frac{\varphi(t)}{t-x} \, dt = f(x), \quad 0 < x < 1, \tag{3.4.24}$$

whose solution is obtained as

$$\varphi(x) = -\frac{1}{\pi^2} \left(\frac{x}{1-x} \right)^{1/2} \int_0^1 \left(\frac{1-t}{t} \right)^{1/2} \left(\frac{f(t)}{t-x} \right) \, dt, \quad 0 < x < 1. \tag{3.4.25}$$

We thus observe that the integral equation (3.4.24) possesses the unique solution as given by (3.4.25) in the special circumstances when both φ and f are bounded at one end $(x = 0)$ and they can be unbounded (with an integrable singularity) at the other end $(x = 1)$. The solution (3.4.25) of the Cauchy type singular integral equation (3.4.24) has already been obtained in section 2.3 by using an elementary method.

2. There are *two* other *important* cases of the integral equation (3.4.1) in the special situation when $\Gamma = (0,1)$, which are as follows:

Case (i): $f(x)$ and $\varphi(x)$ are *unbounded* at both the ends $x = 0$ and $x = 1$.

Case (ii): $f(x)$ and $\varphi(x)$ are *bounded* at both the ends $x = 0$ and $x = 1$.

These two special cases *(i)* and *(ii)* can be handled by choosing special functions $\Phi_0(z)$ of the homogeneous Riemann-Hilbert problem (3.4.11), and we find that the two possible choices are as follows:

For *case (i)*

$$\Phi_0(z) = \Phi_0^{(1)}(z) \equiv z^{\alpha-1}(z-1)^{-\alpha},$$

and for *case (ii)*

$$\Phi_0(z) = \Phi_0^{(2)}(z) = z^{\alpha}(z-1)^{1-\alpha}, \tag{3.4.26}$$

where α is as given by the relation (3.4.16). We find that in *case (i)* we must select $E(z) = $ a constant $= c_0$, say, and in *case (ii)* we must not only select $E(z)$ but also we must have the fact that the forcing function $f(t)$ satisfies the condition

$$\int_0^1 \frac{f(t)}{\Phi_0^{(2)+}(t)} \, dt = 0,$$

which is equivalent to

$$\int_0^1 \frac{f(t)}{t^{\alpha}(1-t)^{1-\alpha}} \, dt = 0. \tag{3.4.27}$$

This condition (3.4.27) is called the *solvability condition* for the given integral equation in the *case (ii)*

The particular limiting case $\rho = 0$ giving rise to the first kind singular integral equation (3.4.24) then produces the following solutions in the above two cases *(i)* and *(ii)*:

In *case (i)* the solution of the integral equation (3.4.24) is given by

$$\varphi(x) = \frac{1}{\{x(1-x)\}^{1/2}} \left[c_0 - \frac{1}{\pi^2} \int_0^1 \frac{\{t(1-t)\}^{1/2} f(t)}{t-x} \, dt \right], \quad 0 < x < 1 \tag{3.4.28}$$

where c_0 is an arbitrary constant.

In *case (ii)* the solution of (3.4.24) is given by

$$\varphi(x) = -\frac{\{x(1-x)\}^{1/2}}{\pi^2} \int_0^1 \frac{f(t)}{\{t(1-t)\}^{1/2}} dt, \quad 0 < x < 1, \quad (3.4.29)$$

provided the forcing function $f(x)$ satisfies the *solvability condition* (see (3.4.27))

$$\int_0^1 \frac{f(t)}{\{t(1-t)\}^{1/2}} dt = 0. \quad (3.4.30)$$

These results have already been obtained in section 2.3 by using an elementary method.

We emphasize that the singular integral equations arising in the crack problem and the surface water wave problem considered in Chapter 1 can now be tackled completely.

3.5 GENERALISED ABEL INTEGRAL EQUATIONS

The generalized Abel integral equation

$$a(x) \int_\alpha^x \frac{\varphi(t)}{(x-t)^\mu} dt + b(x) \int_x^\beta \frac{\varphi(t)}{(t-x)^\mu} = f(x), \quad 0 < \mu < 1, \quad x \in [\alpha, \beta], \quad (3.5.1)$$

was solved by Gakhov (1966, p 531) assuming that $a(x), b(x)$ satisfy Hölder condition on $[\alpha, \beta]$ and $f(x), \varphi(x)$ are such that

$$f(x) = \{(x-\alpha)(\beta-x)\}^\varepsilon f^*(x), \quad \varphi(x) = \frac{\varphi^*(x)}{\{(x-\alpha)(\beta-x)\}^{1-\mu-\varepsilon}} \quad (3.5.2)$$

where $\varepsilon > 0$, $f^*(x)$ possesses Hölder continuous derivative in $[\alpha, \beta]$ and $\varphi^*(x)$ satisfies Hölder condition on $[\alpha, \beta]$. The solution method involved solving a Riemann-Hilbert problem for the determination of a function $\psi(z)$ $(z = x + iy)$ defined by

$$\psi(z) = \frac{1}{R(z)} \int_\alpha^\beta \frac{\varphi(t)}{(t-z)^\mu} dt \quad (3.5.3)$$

where

$$R(z) = \{(z-\alpha)(\beta-z)\}^{\frac{1-\mu}{2}}. \quad (3.5.4)$$

The function $\psi(z)$ is sectionally analytic in the complex z-plane cut along the segment $[\alpha, \beta]$ on the real axis. It may be noted that

$$|\psi(z)| = 0\left(\frac{1}{|z|}\right) \quad \text{as} \ z \to \infty \tag{3.5.5}$$

The RHP was solved by utilizing the Plemelj formulae involving Cauchy singular integrals.

As a particular example of the equation (3.5.1), the solution of the integral equation

$$\int_{\alpha}^{\beta} \frac{\varphi(t)}{|x-t|^{\mu}} \, dt = f(x), \quad 0 < \mu < 1, \ x \in [\alpha, \beta] \tag{3.5.6}$$

was obtained as

$$\varphi(x) = \frac{\sin \pi \mu}{\pi} \frac{d}{dx} \left[\int_{\alpha}^{x} \frac{g(t)}{(x-t)^{1-\mu}} \, dt \right] \tag{3.5.7}$$

where

$$g(x) = \frac{1}{2} f(x) - \frac{\cot\left(\frac{1}{2}\mu\pi\right)}{2\pi} R(x) \int_{\alpha}^{\beta} \frac{f(t)}{R(t)(t-x)} dt, \ x \in [\alpha, \beta] \tag{3.5.8}$$

where the integral is in the sense of CPV.

Remark

The kernel of the integral equation (3.5.6) is *weakly* singular. However the solution, as given by the expressions (3.5.7) and (3.5.8), requires the evaluation of a *strongly* singular integral. Thus the above method has the disadvantage that the solution of a *weakly* singular integral (Abel type singularity) equation is obtained in terms of *strongly* singular integral (Cauchy type singularity).

A straightforward and direct method to solve the integral equation (3.5.1) resulting in solution involving only *weakly* singular integrals of the Abel type has been given by Chakrabarti (2008). This method is now described below.

The method

Let

$$\Phi(z) = \int_\alpha^\beta \frac{\varphi(t)}{(t-z)^\mu}\, dt, \quad 0 < \mu < 1, \quad (z = x + iy). \qquad (3.5.9)$$

Then $\Phi(z)$ is sectionally analytic in the complex z-plane cut along the segment $[\alpha, \beta]$ on the real axis. It is easy to see that, for $x \in [\alpha, \beta]$,

$$\Phi^\pm(x) \equiv \lim_{z \to \pm 0} \Phi(z) = e^{\pm i\pi\mu}\, (A_1\varphi)(x) + (A_2\varphi)(x) \qquad (3.5.10)$$

where the operators A_1, A_2 are defined by

$$(A_1\varphi)(x) = \int_\alpha^x \frac{\varphi(t)}{(x-t)^\mu}\, dt, \quad (A_2\varphi)(x) = \int_x^\beta \frac{\varphi(t)}{(t-x)^\mu}\, dt. \qquad (3.5.11)$$

Using the relations (3.5.10), we find

$$(A_1\varphi)(x) = \frac{\Phi^+(x) - \Phi^-(x)}{2i\,\sin\pi\mu}, \quad (A_2\varphi)(x) = \frac{e^{-i\pi\mu}\Phi^+(x) - e^{i\pi\mu}\Phi^-(x)}{2i\,\sin\pi\mu}, \quad x \in [\alpha, \beta]. \quad (3.5.12)$$

Using these in the given integral equation (3.5.1) we obtain

$$\{a(x) - e^{-i\pi\mu}b(x)\}\Phi^+(x) - \{a(x) - e^{i\pi\mu}b(x)\}\Phi^-(x) = 2i\,\sin\pi\mu\, f(x), \quad x \in [\alpha, \beta]. \quad (3.5.13)$$

The relation (3.5.13) represents the *special Riemann-Hilbert* type problem given by

$$\Phi^+(x) + G(x)\Phi^-(x) = g(x), \quad x \in [\alpha, \beta] \qquad (3.5.14)$$

where

$$G(x) = -\frac{a(x) - e^{i\pi\mu}b(x)}{a(x) - e^{-i\pi\mu}b(x)} = -\exp\left[-2i\,\tan^{-1}\left\{\frac{b(x)\,\sin\pi\mu}{a(x) - b(x)\,\cos\pi\mu}\right\}\right] \qquad (3.5.15)$$

and

$$g(x) = \frac{2i\,\sin\pi\mu}{a(x) - e^{-i\pi\mu}b(x)}\, f(x). \qquad (3.5.16)$$

To solve the special Riemann-Hilbert type problem (3.5.14), we note that

$$|\Phi(z)| = 0\left(\frac{1}{|z|^{\mu}}\right) \quad \text{as} \quad |z| \to \infty. \tag{3.5.17}$$

Let us assume that the homogeneous problem

$$\Phi_0^+(x) + G(x)\Phi_0^-(x) = 0, \quad x \in [\alpha, \beta] \tag{3.5.18}$$

has the solution

$$\Phi_0(z) = e^{\Psi_0(z)}$$

where

$$\Psi_0^+(x) - \Psi_0^-(x) = G_0(x) \tag{3.5.19}$$

with

$$e^{G_0(x)} = -G(x). \tag{3.5.20}$$

To find $\Psi_0(z)$, we utilize the first relation of (3.5.12) in (3.5.19). Thus we can express $\Psi_0(z)$ as

$$\Psi_0(z) = \int_{\alpha}^{\beta} \frac{\psi_0(t)}{(t-z)^{\mu}} \, dt \tag{3.5.21}$$

where

$$\psi_0(x) = \frac{1}{2i \, \sin\pi\mu} \, (A_1^{-1}G_0)(x) \tag{3.5.22}$$

with

$$\left(A_1^{-1}G_0\right)(x) = \frac{\sin\pi\mu}{n} \frac{d}{dx} \int_{\alpha}^{x} \frac{G_0(t)}{(x-t)^{1-\mu}} \, dt. \tag{3.5.23}$$

Using (3.5.15) in (3.5.14) we obtain

$$\frac{\Phi^+(x)}{\Phi_0^+(x)} - \frac{\Phi^-(x)}{\Phi_0^-(x)} = \frac{g(x)}{\Phi_0^+(x)}, \quad x \in [\alpha, \beta] \tag{3.5.24}$$

where

$$\Phi_0^{\pm}(x) = \exp\left[\Psi_0^{\pm}(x)\right]. \tag{3.5.25}$$

We note that $\Psi_0^{\pm}(x)$ can be obtained by using the results in (3.5.10) for the function $\psi_0(z)$ defined by (3.5.21).

Utilizing the first of the formulae (3.5.10), we determine the solution of the Riemann-Hilbert type problem (3.5.24), as given by

$$\frac{\Phi(z)}{\Phi_0(z)} = \int_{\alpha}^{\beta} \frac{\lambda(t)}{(t-z)^{\mu}} dt \qquad (3.5.26)$$

where

$$\lambda(x) = \frac{1}{2\pi i} \frac{d}{dx} \left[\int_{\alpha}^{x} \frac{g(t)}{\Phi_0^{+}(t)(x-t)^{1-\mu}} dt \right]. \qquad (3.5.27)$$

Writing

$$p(t) = \frac{g(t)}{\Phi_0^{+}(t)} \qquad (3.5.28)$$

we see that $\lambda(x)$ in (3.5.27) can be written as

$$\lambda(x) = \frac{1}{2\pi i} \left[\frac{p(\alpha)}{(x-\alpha)^{1-\mu}} + \int_{\alpha}^{x} \frac{p'(t)}{(x-t)^{1-\mu}} dt \right] \qquad (3.5.29)$$

assuming that the derivative $p'(t)$ of the function $p(t)$ exists for $t \in [\alpha, \beta]$.

Now from (3.5.26), we find that

$$\Phi^{\pm}(x) = \Phi_0^{\pm}(x) \left[e^{\pm i\pi\mu} (A_1\lambda)(x) + (A_2\lambda)(x) \right], \quad x \in [\alpha, \beta] \qquad (3.5.30)$$

giving

$$\Phi^{+}(x) - \Phi^{-}(x) = h(x), \quad \text{say}, \quad x \in [\alpha, \beta] \qquad (3.5.31)$$

where

$$h(x) = \left\{ e^{i\pi\mu} \Phi_0^{+}(x) - e^{-i\pi\mu} \Phi_0^{-}(x) \right\}(A_1\lambda)(x)$$
$$+ \left\{ \Phi_0^{+}(x) - \Phi_0^{-}(x) \right\}(A_2\lambda)(x), \quad x \in [\alpha, \beta]. \qquad (3.5.32)$$

By using the first formula in (3.5.12), we obtain from (3.5.31), the solution of the integral equation (3.5.1) as

$$\varphi(x) = \frac{1}{2\pi i} \frac{d}{dx} \left[\int_{\alpha}^{x} \frac{h(t)}{(x-t)^{1-\mu}} dt \right], \quad x \in [\alpha, \beta]. \qquad (3.5.33)$$

This can also be expressed in the equivalent form

$$\varphi(x) = \frac{1}{2\pi i} \left[\frac{h(\alpha)}{(x-\alpha)^{1-\mu}} + \int_{\alpha}^{x} \frac{h'(t)}{(x-t)^{1-\mu}} dt \right], \quad x \in [\alpha, \beta], \qquad (3.5.34)$$

assuming that the derivative $h'(t)$ of $h(t)$ exists for $t \in [\alpha, \beta]$.

An alternative form of the solution of the integral equation (3.5.1) can be derived. This is explained below:

To solve the Riemann-Hilbert type problem (3.5.14), we consider the homogeneous problem

$$\Omega_0^+(x) + e^{-2i\pi\mu} \, \Omega_0^-(x) = 0, \quad x \in [\alpha, \beta] \qquad (3.5.35)$$

instead of the homogeneous problem (3.5.18), we find the alternative representation for $\Phi(z)$ as

$$\Phi(z) = \Omega_0(z) \int_{\alpha}^{\beta} \frac{\omega(t)}{(t-z)^{\mu}} dt \qquad (3.5.36)$$

where

$$\omega(x) = \frac{e^{-i\pi\mu}}{2\pi i} \frac{d}{dx} \int_{x}^{\beta} \frac{g(t)}{\Omega_0^+(t)(t-x)^{1-\mu}}. \qquad (3.5.37)$$

Writing

$$q(t) = \frac{g(t)}{\Omega_0^+(t)} \qquad (3.5.38)$$

we see that $\omega(x)$ in (3.5.37) can be written as

$$\omega(x) = \frac{e^{-i\pi\mu}}{2\pi i} \left[\frac{q(\beta)}{(\beta-x)^{\mu-1}} - \int_{x}^{\beta} \frac{q'(t)}{(t-x)^{1-\mu}} dt \right] \qquad (3.5.39)$$

assuming that the derivative of the function $q(t)$ exists for $t \in [\alpha, \beta]$.

Then, using the limiting values $\Phi^\pm(x)$ of the function $\Phi(z)$ given in (3.5.36), along with the second formula in (3.5.12), we obtain an alternative representation of $\varphi(x)$ as given by

$$\varphi(x) = \frac{1}{2\pi i} \frac{d}{dx} \left[\int_x^\beta \frac{k(t)}{(t-x)^{1-\mu}} \, dt \right], \quad x \in [\alpha, \beta] \qquad (3.5.40)$$

where

$$k(x) = e^{-i\pi\mu} \, \Phi^+(x) - e^{i\pi\mu} \, \Phi^-(x)$$
$$= \left\{ \Omega_0^+(x) - \Omega_0^-(x) \right\} (A_1\omega)(x) + \left\{ e^{-i\pi\mu} \, \Omega_0^+(x) - e^{i\pi\mu} \, \Omega_0^- \right\} (A_2\omega)(x). \qquad (3.5.41)$$

It may be noted that, the formula

$$(A_2^{-1}k)(x) = -\frac{\sin \pi\mu}{\pi} \frac{d}{dx} \left[\int_x^\beta \frac{k(t)}{(t-x)^{1-\mu}} \, dt \right], \quad x \in [\alpha, \beta] \qquad (3.5.42)$$

has been used in obtaining the form (3.5.40). The result (3.5.40) can also be expressed in the equivalent form

$$\varphi(x) = -\frac{1}{2\pi i} \left[\frac{k(\beta)}{(\beta - x)^{1-\mu}} - \int_x^\beta \frac{k'(t)}{(t-x)^{1-\mu}} \, dt \right], \quad x \in [\alpha, \beta], \qquad (3.5.43)$$

whenever the derivative $k'(t)$ of $k(t)$ exists for $t \in [\alpha, \beta]$.

Note: 1. The solution of the integral equation (3.5.1) has been obtained in terms of two forms given by (3.5.33) (or (3.5.34)) and (3.5.40) (or (3.5.43)). These two forms are equivalent although it is not easy to show their equivalence directly.
2. When either $a = 0, b = 1$ or $a = 1, b = 0$, we get back the known solutions of Abel integral equations, using the formula (3.5.33) or (3.5.40).
3. No Cauchy type singular integral equation occurs in the analysis employed here.

3.6 SINGULAR INTEGRAL EQUATIONS WITH LOGARITHMIC KERNEL

Many boundary value problems of mathematical physics give rise to first kind integral equations possessing logarithmically singular kernels of the types

$$\int_{\alpha}^{\beta} \varphi(t) \ln |t - x| \, dt = f(x), \quad \alpha < x < \beta \tag{3.6.1}$$

and

$$\int_{\alpha}^{\beta} \varphi(t) \ln \left| \frac{t - x}{t + x} \right| \, dt = f(x), \quad \alpha < x < \beta \tag{3.6.2}$$

where $\varphi(x)$ and $f(x)$ appearing in (3.6.1) and (3.6.2) are assumed to be differentiable in (α, β).

There exists a number of methods to solve the integral equations (3.6.1) and (3.6.2) which ultimately cast the solution involving Cauchy principal value integrals of the form

$$\int_{\alpha}^{\beta} \frac{\psi(t)}{t - x} \, dt, \quad \alpha < x < \beta \tag{3.6.3}$$

which are strongly singular integrals compared to the weakly singular integrals occurring in (3.6.1) and (3.6.2). For example, the integral equation (3.6.1) has a solution of the form (cf. Porter (1972), Chakrabarti (1997))

$$\varphi(x) = \frac{1}{\pi^2 \{(x-\alpha)(\beta-x)\}^{1/2}} \left[\int_{\alpha}^{\beta} \frac{\{(t-\alpha)(\beta-t)\}^{1/2} f'(t)}{t - x} \, dt \right.$$

$$\left. + \frac{1}{\ln\left(\dfrac{\beta-\alpha}{4}\right)} \int_{\alpha}^{\beta} \frac{f(t)}{\{(t-\alpha)(\beta-t)\}^{1/2}} \, dt \right] \tag{3.6.4}$$

in the case when $\beta - \alpha \neq 4$, and

$$\varphi(x) = \frac{1}{\pi^2 \{(x-\alpha)(\beta-x)\}^{1/2}} \left[\int_{\alpha}^{\beta} \frac{\{(t-\alpha)(\beta-t)\}^{1/2} f'(t)}{t - x} \, dt + C \right] \tag{3.6.5}$$

where C is an arbitrary constant, in the case when $\beta - \alpha = 4$, provided that $f(x)$ satisfies the solvability condition

$$\int_{\alpha}^{\beta} \frac{f(t)}{\{(t-\alpha)(\beta-t)\}^{1/2}} \, dt = 0. \tag{3.6.6}$$

The solutions (3.6.4) and (3.6.5) involve singular integrals of the type (3.6.3).

Chakrabarti (2006) developed direct methods based on complex variable theory to solve the two weakly singular integral equations (3.6.1) and (3.6.2) and obtained solutions which do not involve integrals having stronger singularities of Cauchy type. These methods are described below

The method of solution for the integral equation (3.6.1):

Let

$$\Phi(z) = \int_{\alpha}^{\beta} \varphi(t) \, ln \left(\frac{t-z}{\alpha-z} \right) - A \, ln \left(\frac{\alpha-z}{\beta-z} \right), \quad (z = x+iy) \quad (3.6.7)$$

where A is an arbitrary complex constant. Then the complex valued function is analytic in the complex z-plane cut along the segment $[\alpha, \beta]$ on the real axis. It is easy to observe that

$$\Phi^{\pm}(x) = \int_{\alpha}^{\beta} \varphi_1(t) \, ln \, |t-x| \, dt \mp \pi i \int_{\alpha}^{x} \varphi(t) dt - B \, ln \, (x-\alpha)$$

$$\pm \pi i \, B - A \, ln \left(\frac{x-\alpha}{\beta-x} \right) \pm \pi i \, A \quad \text{for } x \in (\alpha, \beta) \tag{3.6.8}$$

where

$$B = \int_{\alpha}^{\beta} \varphi(t) \, dt. \tag{3.6.9}$$

To derive the result (3.6.8), uses of the following limiting values of the logarithmic functions have been made.

For $x \in (\alpha, \beta)$

$$\left. \begin{array}{l} ln \, (\alpha - z) \to ln \, (x-\alpha) \mp \pi i \text{ as } y \to \pm 0, \\ ln \, (\beta - z) \to ln \, (\beta - x) \text{ as } y \to \pm 0. \end{array} \right\} \tag{3.6.10}$$

The pair of formulae (3.6.8) can be expressed as the following equivalent pair

$$\Phi^+(x) - \Phi^-(x) = 2\pi i \left\{ A + \int_x^\beta \varphi(t)\, dt \right\}, \quad x \in (\alpha, \beta), \quad (3.6.11a)$$

and

$$\Phi^+(x) + \Phi^-(x) = 2 \int_\alpha^\beta \varphi(t)\, ln\, |x - t|\, dt \;-\; 2A\, ln\left(\frac{x - \alpha}{\beta - x}\right) \qquad (3.6.11b)$$
$$-\; 2B\, ln\, (x - \alpha), \quad x \in (\alpha, \beta).$$

We note that by writing

$$r(x) = 2\pi i \left[A + \int_x^\beta \varphi(t)\, dt \right], \quad x \in (\alpha, \beta), \qquad (3.6.12)$$

we obtain

$$\varphi(x) = -\frac{1}{2\pi i}\, r'(x) \qquad (3.6.13)$$

with

$$r(\alpha) = 2\pi i (A + B), r(\beta) = 2\pi i A \qquad (3.6.14)$$

so that the formula (3.6.11a) is expressed as

$$\Phi^+(x) - \Phi^-(x) = r(x), \quad x \in (\alpha, \beta). \qquad (3.6.15)$$

Using the integral equation (3.6.1), we find that the formula (3.6.11b) gives

$$\Phi^+(x) + \Phi^-(x) = f_1(x), \quad x \in (\alpha, \beta). \qquad (3.6.16)$$

where

$$f_1(x) = 2 \left[f(x) - A\, ln\left(\frac{x - a}{\beta - x}\right) - B\, ln\, (x - a) \right], \quad x \in (\alpha, \beta). \quad (3.6.17)$$

The relation (3.6.16) can be regarded as a special case of the more general Riemann-Hilbert problem

$$\Phi^+(x) + g(x)\Phi^-(x) = f(x), \quad x \in (\alpha, \beta) \qquad (3.6.18)$$

involving a sectionally analytic function cut along the segment $[\alpha, \beta]$ on the real axis.

By using the formulae (3.6.11)–(3.6.14), we can easily obtain the solution of the Riemann-Hilbert problem (3.6.16). This is given by

$$\Phi(z) = \Phi_0(z)\left[-\frac{1}{2\pi i} \int_\alpha^\beta \lambda'(t) \, ln \, (t-z) dt - C \, ln \, (\alpha - z) - D \, ln \left(\frac{\alpha - z}{\beta - z} \right) + E(z) \right] \quad (3.6.19)$$

where $\Phi_0(z)$ represents a solution of the homogeneous problem (3.6.16) satisfying

$$\Phi_0^+(x) + \Phi_0^-(x) = 0, \quad x \in (\alpha, \beta), \quad (3.6.20)$$

$$\lambda(x) = \frac{f(x)}{\Phi_0^+(x)}, \quad C = \frac{\lambda(\alpha) - \lambda(\beta)}{2\pi i}, \quad D = \frac{\lambda(\beta)}{2\pi i} \quad (3.6.21)$$

and $E(z)$ represents an entire function in the complex z-plane.

Using the formula (3.6.11a), we then obtain

$$2\pi i \left[\int_x^\beta \varphi(t) \, dt + A \right] = \Phi^+(x) - \Phi^-(x), \quad \alpha < x < \beta$$

$$= \Phi_0^+(x) \, Q^+(x) - \Phi_0^-(x) \, Q^-(x), \quad \alpha < x < \beta \quad (3.6.22)$$

$$= \Phi_0^+(x) \left\{ Q^+(x) + Q^-(x) \right\}, \quad \alpha < x < \beta,$$

where $Q(z)$ represents the term in the square bracket of the expression in the right side of (3.6.19).

Again, using the formula (3.6.11b), applicable to the function $Q(z)$, we obtain

$$2\pi i \left[\int_x^\beta \varphi(t) \, dt + A \right] = \Phi_0^+(x)\left[-\frac{1}{\pi i} \int_\alpha^\beta \lambda'(t) \, ln \, |t - x| \, dt \right.$$

$$\left. - 2C \, ln \, (x - \alpha) - 2D \, ln \left(\frac{x - \alpha}{\beta - x} \right) + 2E(x) \right], \quad \alpha < x < \beta. \quad (3.6.23)$$

We can choose the solution of the homogeneous problem (3.6.20) to be written in the form

$$\Phi_0(z) = \frac{1}{\{(z - \alpha)(z - \beta)\}^{1/2}} \quad (3.6.24)$$

so that

$$\Phi_0^+(x) = \frac{i}{\{(x-\alpha)(\beta - x)\}^{1/2}}. \tag{3.2.25}$$

From the definition of $\Phi(z)$ given by (3.6.7), we find that

$$|\Phi(z)| = 0\left(\frac{1}{|z|}\right) \quad \text{as} \quad |z| \to \infty$$

and since, from (3.6.24),

$$|\Phi_0(z)| = 0\left(\frac{1}{|z|}\right) \quad \text{as} \quad |z| \to \infty,$$

the entire function $E(z)$ appearing in the right side of (3.6.19) must be a constant. Thus choosing $E(z) = \dfrac{D}{2}$, where D is an arbitrary complex constant, we find from (3.6.23),

$$2\pi i\left[\int_x^\beta \varphi(t)\, dt + A\right] = \frac{i}{\{(x-\alpha)(\beta - x)\}^{1/2}}\left[-\frac{1}{\pi i}\int_\alpha^\beta \lambda'(t)\, \ln |t - x|\, dt + D\right]. \tag{3.6.26}$$

where now

$$\lambda(x) = \frac{f_1(x)}{\Phi_0^+(x)} = -2i\left\{(x-\alpha)(\beta - x)\right\}^{1/2}\left[f(x) - B\, \ln (x-\alpha) - A\, \ln\left(\frac{x-\alpha}{\beta - x}\right)\right]. \tag{3.6.27}$$

The relation (3.6.26) eventually determines the solution function $\varphi(x)$, for the equation (3.6.1), by differentiating it with respect to x, and for the purpose of existence of the derivative, which makes the integral in (3.6.1) to converge, we find that we must have

$$\left.\begin{array}{l} D - \dfrac{1}{\pi i}\displaystyle\int_\alpha^\beta \lambda'(t)\, \ln (t-\alpha)\, dt = 0, \\[18pt] D - \dfrac{1}{\pi i}\displaystyle\int_\alpha^\beta \lambda'(t)\, \ln (\beta - t)\, dt = 0. \end{array}\right\} \tag{3.6.28}$$

The two equations in (3.6.28) determine the arbitrary constants D and A in terms of the constant B, which will have to be determined by using the relation (3.6.9), and the process of differentiation applied to (3.6.26), finally produces the solution of the equation (3.6.1) in the form

$$\varphi(x) = -\frac{1}{\pi^2} \frac{d}{dx}\left[\frac{1}{\{(x-\alpha)(\beta-x)\}^{1/2}}\{\frac{\pi}{2}D + \int_\alpha^\beta \frac{d\psi}{dt} \ ln \ |x-t| \, dt\}\right] \quad (3.6.29)$$

where

$$\psi(t) = \{(t-\alpha)(\beta-t)\}^{1/2}\left[f(t) - B \ ln \ (t-\alpha) - A \ ln \left(\frac{t-\alpha}{\beta-t}\right)\right]. \quad (3.6.30)$$

The right side of (3.6.29) can be simplified. It may be noted that

$$\int_\alpha^\beta \frac{d}{dt}[\{(t-\alpha)(\beta-t)\}^{\frac{1}{2}} \ ln\frac{t-\alpha}{\beta-t}] ln \ |x-t| \, dt = -\int_\alpha^\beta \frac{\{(t-\alpha)(\beta-t)\}^{1/2} \ ln \ \frac{t-\alpha}{\beta-t}}{t-x} \, dt \quad (3.6.31)$$

$$= \ -\pi(\beta-\alpha), \quad \alpha < x < \beta$$

and

$$\int_\alpha^\beta \frac{d}{dt}\left[\{(t-\alpha)(\beta-t)\}^{1/2} \ ln \ (t-\alpha)\right] ln \ |x-t| \, dt$$

$$= -\int_\alpha^\beta \frac{\{(t-\alpha)(\beta-t)\}^{1/2} \ ln \ (t-\alpha)}{t-x} \, dt \quad (3.6.32)$$

$$= -\frac{\pi}{2}(\alpha+\beta-2x) \ ln\left(\frac{\beta-\alpha}{4}\right) + \frac{\pi}{2}(\beta-\alpha) + 2\pi \ \{(x-\alpha)(\beta-x)\}^{1/2} \ tan^{-1}\left(\frac{x-a}{b-x}\right)^{1/2}$$

where CPV integrals have been evaluated.

A particular case

Let $f(x) = $ a constant C (say) for the integral equation (3.6.1). Then the solution (3.6.29) reduces to

$$\varphi(x) = \frac{C}{\pi \ ln\left(\dfrac{\beta-\alpha}{4}\right)} \ \{(x-\alpha)(\beta-x)\}^{\frac{1}{2}}. \quad (3.6.33)$$

It is easy to verify that (3.6.33) is indeed the solution of the integral equation

$$\int_\alpha^\beta \varphi(t) \ ln \ |t-x| \, dt = C, \quad \alpha < x < \beta \quad (3.6.34)$$

by using the results

$$\int_{\alpha}^{\beta} \frac{\ln |t-x|}{\{(x-\alpha)(\beta-x)\}^{1/2}} \, dx = \pi \, \ln\left(\frac{\beta-\alpha}{2}\right), \qquad (3.6.35)$$

$$\int_{\alpha}^{\beta} \frac{\ln |t-x|}{\{(t-\alpha)(\beta-t)\}^{1/2}} \, dt = \text{a constant}, \quad a < x < b. \quad (3.6.36)$$

The method of solution for the integral equation (3.6.2):

Let

$$\Phi(z) = \int_{\alpha}^{\beta} \varphi(t) \, \ln \frac{t-z}{t+z} \, dt + A \, \ln \frac{z-\alpha}{z-\beta} + B \, \ln \frac{z+\alpha}{z+\beta} \qquad (3.6.37)$$

where $z = x + iy$, and A, B are complex constants. Then $\Phi(z)$ is sectionally analytic in the complex z-plane cut along segments $(-\beta, -\alpha)$ and (α, β) on the real axis.

Using the results

$$\lim_{y \to \pm 0} \ln z = \begin{cases} \ln x & \text{for } x > 0. \\ \pm i\pi + \ln(-x) & \text{for } x < 0 \end{cases} \qquad (3.6.38)$$

we find that

$$\Phi^{\pm}(x) \equiv \lim_{y \to \pm 0} \Phi(z) = \begin{cases} \mp i\pi \displaystyle\int_{\alpha}^{x} \varphi(t)dt + \int_{\alpha}^{\beta} \varphi(t) \, \ln \left|\frac{t-x}{t+x}\right| \, dt \mp i\pi A \\[2mm] + A \, \ln\left(\frac{x-\alpha}{\beta-x}\right) + B \, \ln \frac{x+\alpha}{\beta+x} \quad \text{for } \alpha < x < \beta, \\[4mm] \pm i\pi \displaystyle\int_{\alpha}^{-x} \varphi(t) \, dt + \int_{\alpha}^{\beta} \varphi(t) \, \ln \left|\frac{t-x}{t+x}\right| \, dt \mp i\pi B \\[2mm] + A \, \ln \left|\frac{x-\alpha}{\beta-x}\right| + B \, \ln \left|\frac{x+\alpha}{\beta+x}\right| \quad \text{for } -\beta < x < -\alpha. \end{cases} \qquad (3.6.39)$$

The equations in (3.6.39) give rise to the following Plemelj-type alternative formulae:

$$\Phi^+(x) - \Phi^-(x) = \begin{cases} -2\pi i \int\limits_{\alpha}^{x} \varphi(t)dt - 2\pi i\, A \text{ for } \alpha < x < \beta, \\ \\ 2\pi i \int\limits_{\alpha}^{-x} \varphi(t)\, dt - 2\pi i\, B \text{ for } -\beta < x < -\alpha \end{cases}$$

(3.6.40)

and

$$\Phi^+(x) + \Phi^-(x) = 2 \int\limits_{\alpha}^{\beta} \varphi(t)\, ln \left|\frac{t-x}{t+x}\right| dt + 2A\, ln \left|\frac{x-\alpha}{\beta-x}\right| + 2B\, ln \left|\frac{x+\alpha}{\beta+x}\right|$$

(3.6.41)

$$\text{for } x \in (-\beta, -\alpha) \cup (\alpha, \beta).$$

By using the relation (3.6.41), we find that the singular integral equation (3.6.2) can be reduced to a Riemann-Hilbert problem, as given by

$$\Phi^+(x) + \Phi^-(x) = 2 \left[f(x) + A\, ln \left|\frac{x-\alpha}{\beta-x}\right| + B\, ln \left|\frac{x+\alpha}{\beta+x}\right| \right]$$

(3.6.42)

$$= 2g(x), \text{ say, for } x \in L \equiv (-\beta, -\alpha) \cup (\alpha, \beta)$$

where

$$g(x) = \begin{cases} f(x) \text{ for } \alpha < x < \beta, \\ -f(-x) \text{ for } -\beta < x < -\alpha \end{cases}$$

(3.6.43)

for the determination of the sectionally analytic function $\Phi(z)$ defined in (3.6.37).

Also by using the right sides of the relation (3.6.40), denoted by $r(x)$, we find that

$$\Phi^+(x) - \Phi^-(x) = r(x) \text{ for } x \in L$$

(3.6.44)

with

$$r'(x) = -2\pi i \begin{cases} \varphi(x) \text{ for } \alpha < x < \beta, \\ -\varphi(-x) \text{ for } -\beta < x < -\alpha. \end{cases}$$

(3.6.45)

Now we observe the following, which will help to solve the Riemann-Hilbert problem (3.6.42).

If we select in (3.6.47)

$$\varphi(x) = 0, \quad A = -B = \frac{1}{2},$$

then the function $\Phi(z)$, as given by (3.6.37) with $\varphi = 0, A = -B = 1/2$) i.e.,

$$\Psi(z) = \frac{1}{2} \, \ln \frac{z-\alpha}{z-\beta} - \frac{1}{2} \, \ln \frac{z+\alpha}{z+\beta} \tag{3.6.46}$$

solves the special RHP as given by (putting $\varphi = 0, A = -B = 1/2$ in (3.6.39))

$$\Psi^{+}(x) - \Psi^{-}(x) = \begin{cases} -\pi i & \text{for } x \in (\alpha, \beta), \\ \pi i & \text{for } x \in (-\beta, -\alpha). \end{cases} \tag{3.6.47}$$

Also if we select in (3.6.37)

$$A = B = 0, \quad \varphi(x) = \begin{cases} -\dfrac{1}{2\pi i} \, r'(x) & \text{for } x \in (\alpha, \beta) \\[2mm] \dfrac{1}{2\pi i} \, r'(-x) & \text{for } x \in (-\beta, -\alpha) \end{cases}$$

with $r(\alpha) = r(-\alpha) = 0$, then the RHP as given by (see the relation (3.6.40))

$$\chi^{+}(x) - \chi^{-}(x) = r(x), \quad x \in L \equiv (-\beta, -\alpha) \cup (\alpha, \beta) \tag{3.6.48}$$

has the solution

$$\chi(z) = -\frac{1}{2\pi i} \int_{\alpha}^{\beta} r'(t) \, \ln \frac{t-z}{t+z} \, dt. \tag{3.6.49}$$

We may add arbitrary entire function to each of the solutions (3.6.46) and (3.6.49) of the two special Riemann-Hilbert problems (3.6.47) and (3.6.48) respectively.

Using these ideas, the Riemann-Hilbert problem described by (3.6.41) can be solved and its solution is given by

$$\Phi(z) = \Phi_0(z) \left[-\frac{1}{\pi i} \int_{\alpha}^{\beta} \frac{d}{dt} \left\{ \frac{f(t) + A \, \ln \dfrac{t-\alpha}{\beta+t} + B \, \ln \dfrac{t+\alpha}{\beta+t}}{\Phi_0^{+}(t)} \right\} \ln \frac{z-t}{z+t} \, dt + E(z) \right] \tag{3.6.50}$$

where $\Phi_0(z)$ represents the solution of the homogeneous problem

$$\Phi_0^+(x) + \Phi_0^- = 0, \quad x \in L \equiv (-\beta, -\alpha) \cup (\alpha, \beta) \qquad (3.6.51)$$

and $E(z)$ is an entire function of z. By considering the order of $\Phi(z)$ defined by (3.6.37) for large z, we find that $E(z)$ is a polynomial of degree 2. Now solution of (3.6.51) can be taken as

$$\Phi_0(z) = \left\{(z^2 - \alpha^2)(z^2 - \beta^2)\right\}^{-1/2} \qquad (3.6.52)$$

so that

$$\Phi_0^{\pm}(x) = \pm i \left\{(x^2 - \alpha^2)(\beta^2 - x^2)\right\}^{-1/2}, \quad x \in L. \qquad (3.6.53)$$

The solution of the integral equation (3.6.2), thus, can be finally determined by using the relations (3.1.50), (3.6.52) and (3.6.53) along with (3.6.44), and we find that the solution $\varphi(x)$ can be expressed as

$$\varphi(x) = -\frac{1}{2\pi i}\frac{d}{dx}\left\{\Phi^+(x) - \Phi^-(x)\right\}$$

$$= -\frac{1}{\pi^2}\frac{d}{dx}\left[\frac{\psi(x)}{\left\{(x^2 - a^2)(\beta^2 - x^2)\right\}^{1/2}}\right], \quad x \in (\alpha, \beta) \qquad (3.6.54)$$

where

$$\psi(x) = \int_{\alpha}^{\beta}\frac{d}{dt}\left[\left\{(t^2 - \alpha^2)(\beta^2 - t^2)\right\}^{1/2}\left\{f(t) + A\ln\frac{t - \alpha}{\beta - t} + B\ln\frac{t + \alpha}{\beta + t}\right\}\right]\ln\left|\frac{t - x}{t + x}\right|dt \qquad (3.6.55)$$
$$+ \pi\,(C_1 + C_2 x + Dx^2)$$

where C_1, C_2 are two arbitrary constants, and

$$D = \pi i \int_{\alpha}^{\beta}\varphi(x)dx. \qquad (3.6.56)$$

It may be observed that the form (3.6.54) of the function $\varphi(x)$ can solve the integral equation (3.6.2), if and only if the following four conditions are satisfied

(i) $\psi(-\beta) = 0$, (ii) $\psi(-\alpha) = 0$, (iii) $\psi(\alpha) = 0$, (iv) $\psi(\beta) = 0$. $\qquad$ (3.6.57)

The five conditions, one in (3.6.56) and four in (3.6.57), will determine the five constants D, A, B, C_1 and C_2 and the final form of the solution

$\varphi(x)$ can then be determined completely which does not depend on any arbitrary constant.

It may be noted that the conditions (i) and (ii) of (3.6.57) can be avoided if we choose $A = 0 = B$.

This completes the method of solution of the integral equation (3.6.2). Examples for particular forms of the forcing function can be considered although no such example has been undertaken here.

Remark

The form of the solution (3.6.54) of the integral equation (3.6.2) possesses terms which are unbounded at $x = \alpha$ and $x = \beta$. Hence, in order that the solution is bounded at $x = \alpha$ and $x = \beta$, we must have two solvability conditions to be satisfied by the forcing term, as given by the equation

$$\text{(i) } \psi'(\alpha) = 0 \text{ and (ii) } \psi'(\beta) = 0. \tag{3.6.58}$$

Chakrabarti et al (2003) demonstrated the utility of bounded solution of the integral equation (3.6.2) in connection with the study of surface water wave scattering by a vertical barrier with a gap.

3.7 SINGULAR INTEGRAL EQUATION WITH LOGARITHMIC KERNEL IN DISJOINT INTERVALS

In the previous section two first kind weakly singular integral equations with logarithmic kernel in a single interval have been solved by reducing them to appropriate Riemann-Hilbert problems. In this section a logarithmically singular integral equation in a finite number of disjoint intervals have been considered for solution. The integral equation is solved after reducing it to a Riemann-Hilbert problem. The method as developed by Banerjea and Rakshit (2007) is described below.

We consider the integrral equation

$$\sum_{j=1}^{n} \int_{\alpha_j}^{\beta_j} \varphi(t) \ln |t - x| \, dt = f(x), \quad x \in L \equiv \bigcup_{j=1}^{n} (\alpha_j, \beta_j). \tag{3.7.1}$$

Let

$$\Phi(z) = \sum_{j=1}^{n} \int_{\alpha_j}^{\beta_j} \varphi(t) \ln (t - z) \, dt - A \sum_{j=1}^{n} \ln (\alpha_j - z) - \sum_{j=1}^{n} B_j \ln \frac{\alpha_j - z}{\beta_j - z}, \quad (z = x + iy) \tag{3.7.2}$$

where

$$A = \frac{1}{n} \sum_{j=1}^{n} \int_{\alpha_j}^{\beta_j} \varphi(t)\, dt \tag{3.7.3}$$

and B_j's ($j = 1, 2, ..., n$ are arbitrary. Then $\Phi(z)$ is sectionally analytic in the complex z-plane cut along the segments (α_1, β_1), (α_2, β_2),..., (α_n, β_n) on the real axis.

By using the results in (3.6.48), we find that for $x \in L$,

$$\Phi^{\pm}(x) = \begin{cases} \sum_{j=1}^{n} \int_{\alpha_j}^{\beta_j} \varphi(t)\, ln\, |t-x|\, dt - A\sum_{j=1}^{n} ln\, |\alpha_j - x| - \sum_{j=1}^{n} B_j\, ln\, \left|\frac{\alpha_j - x}{\beta_j - x}\right| \\ \mp \int_{a_1}^{x} \varphi(t)\, dt \pm i\pi\, (A + B_1), \quad \text{for } x \in (\alpha_1, \beta_1), \\[2mm] \sum_{j=1}^{n} \int_{\alpha_j}^{\beta_j} \varphi(t)\, ln\, |t-x|\, dt - A\sum_{j=1}^{n} ln\, |\alpha_j - x| - \sum_{j=1}^{n} B_j\, ln\, \left|\frac{\alpha_j - x}{\beta_j - x}\right| \\ \mp i\pi \sum_{\alpha_j}^{k-1} \int_{\alpha_j}^{\beta_j} \varphi(t)\, dt \mp i\pi \int_{\alpha_j}^{x} \varphi(t)\, dt \pm i\pi(kA + B_k), \quad \text{for } x \in (\alpha_k, \beta_k),\ k = 2, 3, ..., n. \end{cases} \tag{3.7.4}$$

The equations (3.7.4) give rise to the following *Plemelj-type* formula for the sectionally analytic function $\Phi(z)$:

$$\Phi^+(x) + \Phi^-(x) = h(x), \quad \text{for } x \in L, \tag{3.7.5}$$

$$\Phi^+(x) - \Phi^-(x) = r(x), \quad \text{for } x \in L, \tag{3.7.6}$$

where

$$h(x) = 2\left\{ f(x) - A\sum_{j=1}^{n} ln\, |\alpha_j - x| - \sum_{j=1}^{n} B_j\, ln\, \left|\frac{\alpha_j - x}{\beta_j - x}\right| \right\} \tag{3.7.7}$$

and

$$r(x) = \begin{cases} -2\pi i \int_{\alpha_1}^{x} \varphi(t)\, dt + 2\pi i(A + B_1) \quad \text{for } x \in (\alpha_1, \beta_1), \\[2mm] -2\pi i \sum_{j=1}^{k-1} \int_{\alpha_j}^{\beta_j} \varphi(t)\, dt - 2\pi i \int_{\alpha_k}^{x} \varphi(t)\, dt + 2\pi i\, (kA + B_k) \end{cases} \tag{3.7.8}$$

$$\text{for } x \in (\alpha_k, \beta_k),\ k = 2.3..., n.$$

Thus from (3.7.8) we find that

$$\varphi(x) = -\frac{1}{2\pi i}\, r'(x), \quad x \in (\alpha_k, \beta_k), \quad k = 1, 2, ..., n. \qquad (3.7.9)$$

and in the expression for $r(x)$ given by (3.7.8), the constants A and B_j $(j = 1, 2, ...n)$ are to be determined such that

$$A = \frac{1}{2n\pi i} \sum_{j=1}^{n} \left\{ r(\alpha_j) - r(\beta_j) \right\},$$

$$B_1 = -A + \frac{1}{2\pi i}\, r(\alpha_1), \qquad (3.7.10)$$

$$B_k = -kA + \frac{1}{2\pi i} \left[\sum_{j=1}^{k} r(\alpha_j) - \sum_{j=1}^{k-1} r(\beta_j) \right], \quad k=2,3,...n.$$

A solution of the homogeneous problem

$$\Phi_0^+(x) + \Phi_0^-(x) = 0, \quad x \in (\alpha_j, \beta_j), \quad j = 1, 2, ...n \qquad (3.7.11)$$

is given by

$$\Phi_0(z) = R(z) \equiv \prod_{j=1}^{n} \left\{ (z - \alpha_j)(z - \beta_j) \right\}^{\frac{1}{2}}, \qquad (3.7.12)$$

then

$$\Phi_0^{\pm}(x) = \begin{cases} \pm \dfrac{(-1)^n i}{|R(x)|}, & x \in (\alpha_1, \beta_1), \\[2mm] \pm \dfrac{(-1)^{n-1} i}{|R(x)|}, & x \in (\alpha_2, \beta_2), \\[2mm] \cdots\cdots\cdots\cdots\cdots \\[2mm] \pm \dfrac{-i}{|R(x)|}, & x \in (\alpha_n, \beta_n). \end{cases} \qquad (3.7.13)$$

Thus, following a somewhat similar argument in section 3.6, we find that the Riemann-Hilbert problem described by (3.7.5) has the solution

$$\Phi(z) = \Phi_0(z) \left[-\frac{1}{2\pi i} \sum_{j=1}^{n} \int_{\alpha_j}^{\beta_j} g'(t) \, ln \, (t-z) \, dt - C \sum_{j=1}^{n} ln \, (\alpha_j - z) \right.$$

$$\left. - \sum_{j=1}^{n} D_j \, ln \, \frac{\alpha_j - z}{\beta_j - z} + E(z) \right] \qquad (3.7.14)$$

where

$$g(x) = \frac{h(x)}{\Phi_0^+(x)},$$

$$C = -\frac{1}{2\pi i n} \sum_{j=1}^{n} \left\{ g(\beta_j) - g(\alpha_j) \right\},$$

$$D_1 = -C + \frac{1}{2\pi i} g(\alpha_1), \qquad (3.7.15)$$

$$D_k = -kC + \frac{1}{2\pi i} \left[\sum_{j=1}^{k} g(\alpha_j) - \sum_{j=1}^{k-1} g(\beta_j) \right], \quad k = 2, 3 \ldots n,$$

and $E(z)$ is an entire function of z in the form of a polynomial of degree $(n-1)$.

It may be noted here that after substituting $\Phi_0^+(x)$ from (3.7.13) and (3.7.12) into $g(x)$ in (3.7.15), one gets

$$g(\alpha_i) = g(\beta_j) = 0, \quad j = 1, 2 \ldots n.$$

Hence

$$C = 0, \quad D_1 = 0, \quad D_k = 0, \quad k = 2, 3 \ldots n. \qquad (3.7.16)$$

Thus from (3.7.14) we get

$$\Phi(z) = \Phi_0(z) \left[-\frac{1}{2\pi i} \sum_{j=1}^{n} \int_{\alpha_j}^{\beta_j} g'(t) \, ln \, (t-z) \, dt + E(z) \right]. \qquad (3.7.17)$$

Using (3.7.17) in (3.7.6) we obtain for $x \in (\alpha_j, \beta_j)$,

$$r(x) = \frac{(-1)^{n-j+1} i}{|R(x)|} \left[-\frac{1}{\pi i} \sum_{j=1}^{n} \int_{\alpha_j}^{\beta_j} g'(t) \, ln \, |t-x| \, dt + 2E(x) \right]. \qquad (3.7.18)$$

Substituting $h(x)$ from (3.7.7) and $\Phi_0^+(x)$ from (3.7.13) into $g(x)$ in (3.7.15) we get

$$g(x) = 2i(-1)^{n-j+2}\psi(x), \quad x \in (\alpha_j, \beta_j), \quad j \geq 1$$

where

$$\psi(x) = |R(x)| \left[f(x) - A \sum_{j=1}^{n} \ln |\alpha_j - x| - \sum_{j=1}^{n} B_j \ln \left| \frac{\alpha_j - x}{\beta_j - x} \right| \right]. \qquad (3.7.19)$$

Finally substituting $r(x)$ from (3.7.18), (3.7.19) into (3.7.9) we obtain $\varphi(x)$ for $x \in (\alpha_j, \beta_j)$ as

$$\varphi(x) = -\frac{1}{\pi^2} \frac{d}{dx} \left[\frac{1}{|R(x)|} \left\{ \sum_{j=1}^{n} \int_{\alpha_j}^{\beta_j} \frac{d\psi}{dt} \ln|t - x| \, dt + (-1)^j \pi E(x) \right\} \right], \qquad (3.7.20)$$

where $\pi E(x)$ is a polynomial of degree $n-1$ given by

$$\pi E(x) = d_1 + d_2 x + \cdots + d_n x^{n-1}. \qquad (3.7.21)$$

Now the original integral equation (3.7.1) can be solved if the following $2n$ consistency conditions are satisfied.

$$\int_{\alpha_j}^{\beta_j} \frac{d}{dt} (\psi(t)) \ln|t - \alpha_j| \, dt + (-1)^j \pi E(\alpha_j) = 0,$$

$$(3.7.22)$$

$$\int_{\alpha_j}^{\beta_j} \frac{d}{dt} (\psi(t)) \ln|t - \beta_j| \, dt + (-1)^j \pi E(\beta_j) = 0, \quad j = 1, 2 \ldots n.$$

These $2n$ conditions determine the $2n$ constants d_k's in (3.7.21) and B_k's in (3.7.19), $k = 1, 2, \ldots, n$.

Chapter 4

Special Methods of Solution of Singular Integral Equations

In this chapter we describe certain special methods of solution of singular integral equations involving logarithmic type and Cauchy type singularities in the kernels. These methods avoid the detailed uses of the complex variable methods described in Chapter 3. The special methods of solution of singular integral equations discussed in this chapter, have been the subject of studies of various researchers such as Estrada and Kanwal (1985a, 1985b, 1987, 2000), Boersma (1978), Brown (1977), Chakrabarti (1995), Chakrabarti and Williams (1980) and others.

4.1 INTEGRAL EQUATIONS WITH LOGARITHMICALLY SINGULAR KERNELS

(a) We first consider the problem of solving the singular integral equation

$$\int_{-1}^{1} \frac{\ln |x-t| \, \varphi(t)}{\left(1-t^2\right)^{1/2}} \, dt = f(x), \quad -1 < x < 1 \qquad (4.1.1)$$

where $f(x)$ and $\varphi(x)$ are differentiable functions in the interval $(-1,1)$ so that the mathematical analysis, as described below, to obtain the solution of the equation (4.1.1), is applicable. It may be noted that the method is also applicable to less restricted class of functions, e.g. f and φ are integrable in $(-1,1)$.

By a straightforward transformation of the variables and the functions involved, the integral equation (4.1.1) can be viewed as an equation holding in the general interval (a,b) where a and b are real numbers and $b > a$.

Before explaining a special method of solution of the equation (4.1.1), which will require the knowledge of the Chebyshev polynomials $T_n(x) = \cos(n \cos^{-1} x)$, $-1 < x < 1$ $(n = 0, 1, ...)$, we mention below a few results which will be needed in the analysis.

(i) The Chebyshev polynomials $T_n(x)$ $(n = 0, 1, ...)$ possess the orthogonality property

$$\int_{-1}^{1} \frac{T_n(x)\, T_m(x)}{\left(1 - x^2\right)^{1/2}}\, dx = \begin{cases} 0, & \text{for } n \neq m, \\ \dfrac{\pi}{2}, & \text{for } n = m \neq 0, \\ \pi, & \text{for } n = m = 0. \end{cases} \tag{4.1.2}$$

(ii) Any function $f(x)$ defined in $[-1, 1]$ and satisfying the condition

$$\int_{-1}^{1} \frac{\left|f(x)\right|^2}{\left(1 - x^2\right)^{1/2}}\, dx < \infty \tag{4.1.3}$$

can be expanded in a Chebyshev series as given by

$$f(x) = \sum_{n=0}^{\infty} c_n\, T_n(x) \tag{4.1.4}$$

where

$$c_0 = \frac{1}{\pi} \int_{-1}^{1} \frac{f(x)}{\left(1 - x^2\right)^{1/2}}\, dx, \quad c_n = \frac{2}{\pi} \int_{-1}^{1} \frac{f(x)\, T_n(x)}{\left(1 - x^2\right)^{1/2}}\, dx \quad (n \geq 1), \tag{4.1.5}$$

with the series (4.1.4) being convergent in mean square sense, with respect to the weight function $(1 - x^2)^{1/2}$.

As a special case of the expansion (4.1.4), we find that, for $-1 \leq x, t \leq 1$, we have

$$ln\, |x - t| = -ln\, 2 - 2 \sum_{n=1}^{\infty} \frac{T_n(x)\, T_n(t)}{n},$$

which can be verified easily with the aid of the following results:

$$\int_{-1}^{1} \frac{ln\, |x - t|}{\left(1 - t^2\right)^{1/2}}\, dt = -\pi\, ln\, 2, \quad -1 < x < 1 \tag{4.1.6}$$

and

$$\int_{-1}^{1} \frac{\ln |x-t|\, T_n(t)}{\left(1-t^2\right)^{1/2}}\, dt = -\frac{\pi}{n}\, T_n(x), \quad -1 < x < 1, \quad n \geq 1. \ (4.1.7)$$

Observation: It is interesting to observe from the relation (4.1.6) that the integral equation

$$\int_{-1}^{1} \frac{\ln |x-t|\, \varphi(t)}{\left(1-t^2\right)^{1/2}}\, dt = 1, \quad -1 < x < 1, \tag{4.1.8}$$

possesses the solution, as given by

$$\varphi(t) = -\frac{1}{\pi \ln 2}, \quad -1 < t < 1. \tag{4.1.9}$$

Let us now explain a special method of solution of the integral equation (4.1.1).

Let us assume that the function $f(x)$ and $\varphi(t)$ appearing in equation (4.1.1) can be expanded in Chebyshev series as given by

$$f(x) = \sum_{n=0}^{\infty} c_n\, T_n(x) \tag{4.1.10}$$

and

$$\varphi(x) = \sum_{n=0}^{\infty} d_n\, T_n(x) \tag{4.1.11}$$

where c_n's are known constants, as given by the formulae (4.1.5), and d_n's are unknown constants to be determined.

By using the expressions (4.1.10) and (4.1.11) into the equation (4.1.1) and utilizing the results (4.1.6) and (4.1.7), we obtain

$$(-\pi \ln 2)\, d_0 - \pi \sum_{n=1}^{\infty} \frac{d_n}{n}\, T_n(x) = \sum_{n=0}^{\infty} c_n\, T_n(x), \quad -1 < x < 1. \ (4.1.12)$$

We thus find that the unknown constants d_n's are given by the following relations:

$$d_0 = \frac{c_0}{\pi \, ln \, 2} = -\frac{1}{\pi^2 \, ln \, 2} \int_{-1}^{1} \frac{f(x)}{\left(1-x^2\right)^{1/2}} \, dx,$$

and

$$d_n = -\frac{n}{\pi} \, c_n = -\frac{2n}{\pi^2} \int_{-1}^{1} \frac{f(x)T_n(x)}{\left(1-x^2\right)^{1/2}} \, dx, \quad n \geq 1. \qquad (4.1.13)$$

The relation (4.1.11), along with the relations (4.1.13), solve the integral equation (4.1.1) completely, for the class of functions $f(x)$ for which the constants d_n, as given by the relations (4.1.13), help the series (4.1.11) to converge in the *mean square* sense described before.

It is easily verified that the solution of the integral equation (4.1.1), in the special case when $f(x) = 1$ (i.e. the integral equation (4.1.8)), is given by

$$\varphi(t) = d_0 \, T_0(x) = -\frac{1}{\pi \, ln \, 2}, \qquad (4.1.14)$$

which was already observed earlier in the relations (4.1.8) and (4.1.9).

Now, denoting by L_0 the operator as defined by

$$(L_0\varphi)(x) \equiv L_0\left(\varphi(t); x\right) = \int_{-1}^{1} \frac{ln \, |x-t| \, \varphi(t)}{\left(1-t^2\right)^{1/2}} \, dt, \qquad (4.1.15)$$

we observe that, the eigenvalues of the operator L_0 are

$$\lambda_n = \begin{cases} -\pi \, ln \, 2, & \text{for } n = 0, \\ -\dfrac{\pi}{n} & \text{for } n \geq 1 \end{cases} \qquad (4.1.16)$$

with the associated eigenfunctions $T_n(x)$ $(n \geq 0)$.

We then consider the integral equation of the second kind as given by

$$\varphi(x) - \lambda(L_0\varphi)\,(x) = f(x), \quad -1 < x < 1. \qquad (4.1.17)$$

We find that if we expand the functions f and φ in terms of Chebyshev series (4.1.10) and (4.1.11), then the solution of the integral equation (4.1.17) can be expressed in the form

$$\varphi(x) = \int_{-1}^{1} R(x, y; \lambda) \ f(y) \ dy \qquad (4.1.18)$$

where the *resolvent* R is given by the formula

$$\left(1 - y^2\right)^{1/2} R(x, y; \lambda) = \frac{1}{\pi \left(1 + \lambda \pi \ \ln 2\right)} + \frac{2}{\pi} \sum_{n=1}^{\infty} \frac{T_n(x) \ T_n(y)}{1 + \dfrac{\lambda \pi}{n}} \qquad (4.1.19)$$

whenever $\lambda \lambda_n \neq -1$, with λ_n's $(n \geq 0)$ being given by the relations (4.1.16).

It is also observed from above that if $\lambda \lambda_n = -1$, then the solution of the integral equation (4.1.17) exists *if an only if*

$$\int_{-1}^{1} \frac{T_n(x) \ f(x)}{\left(1 - x^2\right)^{1/2}} \ dx = 0. \qquad (4.1.20)$$

(b) We next consider the singular integral equation

$$\int_{-1}^{1} \ln |x - t| \ \varphi(t) \ dt = f(x), \quad -1 < x < 1. \qquad (4.1.21)$$

By formally differentiating both sides of the equation (4.1.21) with respect to x, we obtain the singular integral equation with Cauchy type kernel

$$\int_{-1}^{1} \frac{\varphi(t)}{t - x} \ dt = - f'(x), \quad -1 < x < 1. \qquad (4.1.22)$$

Using the results of Chapter 3, we can express the general solution of the singular integral equation (4.1.21) in the form

$$\varphi(x) = \frac{1}{\left(1 - x^2\right)^{1/2}} \left[C + \frac{1}{\pi^2} \int_{-1}^{1} \frac{\left(1 - t^2\right)^{1/2} f'(t)}{t - x} \ dt \right], \quad -1 < x < 1, \qquad (4.1.23)$$

where C is an arbitrary constant.

We easily observe, from equation (4.1.23), that

$$\int_{-1}^{1}\varphi(x)\,dx = C\int_{-1}^{1}\frac{1}{\left(1-x^2\right)^{1/2}}\,dx + \frac{1}{\pi^2}\int_{-1}^{1}(1-t^2)^{1/2}f'(t)\left\{\int_{-1}^{1}\frac{dx}{\left(1-x^2\right)^{1/2}(t-x)}\right\}dt$$

producing

$$C = \frac{1}{\pi}\int_{-1}^{1}\varphi(x)\,dx, \qquad\qquad (4.1.24)$$

after noting that

$$\int_{-1}^{1}\frac{dx}{\left(1-x^2\right)^{1/2}(t-x)} = 0 \ \text{ for } \ -1<t<1.$$

Now from the equation (4.1.21) we derive that

$$\int_{-1}^{1}\frac{1}{\left(1-x^2\right)^{1/2}}\left\{\int_{-1}^{1}\ln|x-t|\,\varphi(t)\,dt\right\}dx = \int_{-1}^{1}\frac{f(x)}{\left(1-x^2\right)^{1/2}}\,dx,$$

giving

$$\int_{-1}^{1}\varphi(t)\left\{\int_{-1}^{1}\frac{\ln|x-t|}{\left(1-x^2\right)^{1/2}}\,dx\right\}dt = \int_{-1}^{1}\frac{f(x)}{\left(1-x^2\right)^{1/2}}\,dx.$$

By using the results (4.1.8) and (4.1.9), this produces

$$-\pi\,\ln 2\int_{-1}^{1}\varphi(t)\,dt = \int_{-1}^{1}\frac{f(x)}{\left(1-x^2\right)^{1/2}}\,dx. \qquad\qquad (4.1.25)$$

The use of the two results (4.1.24) and (4.1.25) suggests, therefore, that if the function $\varphi(x)$, as given by the relation (4.1.23), has to be the solution of the integral equation (4.1.21), the constant C must be given by the relation

$$C = -\frac{1}{\pi^2\,\ln 2}\int_{-1}^{1}\frac{f(x)}{\left(1-x^2\right)^{1/2}}\,dx, \qquad\qquad (4.1.26)$$

and, then, the solution of the singular integral equation (4.1.21) is obtained in the form

$$\varphi(x) = \frac{1}{\pi^2 \left(1-x^2\right)^{1/2}} \left[\int_{-1}^{1} \frac{\left(1-t^2\right)^{1/2} f'(t)}{t-x} \, dx - \frac{1}{\ln 2} \int_{-1}^{1} \frac{f(t)}{\left(1-t^2\right)^{1/2}} \, dt \right], -1 < x < 1. \quad (4.1.27)$$

Remarks

1. The same equation (4.1.22) results even if (4.1.21) is replaced by

$$\int_{-1}^{1} \ln |x-t| \, \varphi(t) \, dt = f(x) + D, \quad -1 < x < 1 \quad (4.1.21')$$

where D is an arbitrary constant. Thus, (4.1.23) will represent the solution of (4.1.21') also (!) with

$$C = \frac{1}{\pi} \int_{-1}^{1} \varphi(t) \, dt.$$

But (4.1.25) changes, as in also the case with (4.1.26), giving

$$C = -\frac{1}{\pi^2 \ln 2} \int_{-1}^{1} \frac{f(t)+D}{\left(1-t^2\right)^{1/2}} \, dt$$

$$= (\text{old } C) - \frac{D}{\pi^2 \ln 2} \int_{-1}^{1} \frac{dt}{\left(1-t^2\right)^{1/2}}$$

$$= (\text{old } C) - \frac{D}{\pi^2 \ln 2}.$$

Thus, an extra solution of (4.1.21), of the form

$$-\frac{D}{\pi \ln 2} \left(\frac{1}{1-x^2}\right)^{1/2}$$

is possible.

2. The solution of $\varphi(x)$, as given by the relation (4.1.27), is unbounded at both end points $x = -1$ and $x = 1$. We must mention here that if $\varphi(x)$ has to be *bounded* at both the end points $x = -1$ and $x = 1$, then for bounded solution of the equation (4.1.22), we must have that

$$\int_{-1}^{1} \frac{f'(t)}{\left(1-t^2\right)^{1/2}}\, dt = 0 \tag{4.1.28}$$

giving the solution of $\varphi(t)$ as

$$\varphi(x) = \frac{\left(1-x^2\right)^{1/2}}{\pi^2} \int_{-1}^{1} \frac{f'(t)}{\left(1-t^2\right)^{1/2}(t-x)}\, dt, \quad -1 < x < 1. \tag{4.1.29}$$

We remark, at this stage, that the result (4.1.27) can be rewritten as

$$\varphi(x) = \frac{1}{\pi^2}\left[\left(1-x^2\right)^{1/2} \int_{-1}^{1} \frac{f'(t)}{\left(1-t^2\right)^{1/2}(t-x)}\, dt - \frac{1}{\left(1-x^2\right)^{1/2}} \int_{-1}^{1} \frac{(t+x)f'(t)}{\left(1-t^2\right)^{1/2}}\, dt \right] \tag{4.1.30}$$

$$- \frac{1}{\pi^2\, ln\, 2} \frac{1}{\left(1-x^2\right)^{1/2}} \int_{-1}^{1} \frac{f(t)}{\left(1-t^2\right)^{1/2}}\, dt.$$

Thus, we notice that, the solutions (4.1.29) and (4.1.30) will agree with each other and produce the bounded solution (4.1.29), at the two ends $x = -1$ and $x = 1$, *if and only if* the following two conditions hold good:

(i)
$$\int_{-1}^{1} \frac{f'(t)}{\left(1-t^2\right)^{1/2}}\, dt = 0 \tag{4.1.31}$$

which is the condition (4.1.28) and

(ii)
$$\int_{-1}^{1} \frac{t\, f'(t)}{\left(1-t^2\right)^{1/2}}\, dt = -\frac{1}{ln\, 2} \int_{-1}^{1} \frac{f(t)}{\left(1-t^2\right)^{1/2}}\, dt. \tag{4.1.32}$$

(c) We now consider the singular integral equation

$$\frac{1}{\pi} \int_{0}^{1} \varphi(t)\, ln\, \frac{t+x}{t-x}\, dt = f(x), \quad 0 < x < 1. \tag{4.1.33}$$

where $f(0) = 0$ for consistency. This integral equation arises in the study of problems on water wave scattering by vertical barriers in the linearised theory of water waves (cf. Chakrabarti et al (2003)).

By differentiating the equation (4.1.33) formally, with respect to x, as was done for the problem in (b), we obtain the Cauchy type singular integral equation

$$\frac{1}{\pi} \int_0^1 \frac{2t\, \varphi(t)}{t^2 - x^2}\, dt = f'(x), \quad 0 < x < 1. \tag{4.1.34}$$

This integral equation can be cast into the new form

$$\frac{1}{\pi} \int_0^1 \frac{\varphi_1(u)}{u - y}\, du = f_1(y), \quad 0 < y < 1 \tag{4.1.35}$$

by using the following transformations

$$t^2 = u, \quad x^2 = y, \quad \varphi_1(u) = \varphi(u^{1/2}), \quad f_1(y) = f'(y^{1/2}). \tag{4.1.36}$$

The general solution of the singular integral equation (4.1.35), which is of the Cauchy type, can be easily written down as (see section 3)

$$\varphi_1(u) = \frac{1}{\pi} \frac{1}{\{u(1-u)\}^{1/2}} \left[c - \int_0^1 \frac{\{y(1-y)\}^{1/2}\, f(y)}{y - u}\, dy \right], \quad 0 < u < 1, \tag{4.1.37}$$

where c is an arbitrary constant. Transforming back to the old variables, we rewrite the relation (4.1.37) as

$$\varphi(t) = \frac{1}{\pi} \frac{1}{t(1-t^2)^{1/2}} \left[c - 2 \int_0^1 \frac{x^2(1-x^2)^{1/2}}{x^2 - t^2}\, f'(x)\, dx \right], \quad 0 < t < 1. \tag{4.1.38}$$

We then notice, by standard integrations, that

$$c = 2 \int_0^1 t\, \varphi(t)\, dt. \tag{4.1.39}$$

We observe from the original integral equation (4.1.33) that

$$\int_0^1 \frac{x f(x)}{\left(1 - x^2\right)^{1/2}}\, dx = \frac{1}{\pi} \int_0^1 \varphi(t) \left\{ \int_0^1 \frac{x}{\left(1 - x^2\right)^{1/2}}\, \ln\left| \frac{x+t}{x-t} \right|\, dx \right\}\, dt \tag{4.1.40}$$

$$= \int_0^1 t\, \varphi(t)\, dt$$

obtained by using the result

$$\int_0^1 \frac{x}{\left(1-x^2\right)^{1/2}} \ln \left|\frac{x+t}{x-t}\right| dx = \pi t, \quad 0 < t < 1. \tag{4.1.41}$$

We then find by using the two relations (4.1.39) and (4.1.40), that we must choose the constant c as given by

$$c = 2 \int_0^1 \frac{x f(x)}{\left(1-x^2\right)^{1/2}} dx. \tag{4.1.42}$$

Thus, with c being chosen to be given by the relation (4.1.42), the relation (4.1.38) produces the general solution of the singular integral equation (4.1.33), which does not involve any arbitrary constant, and this general solution has the property that it is unbounded at both the end points $t = 0$ and $t = 1$.

Now, we can formally rewrite the solution (4.1.38) as

$$\varphi(t) = \frac{1}{\pi} \frac{1}{t\left(1-t^2\right)^{1/2}} \left[c - 2(1-t^2) \left\{ \int_0^1 \frac{f'(x)}{\left(1-x^2\right)^{1/2}} dx + t^2 \int_0^1 \frac{f'(x)}{\left(1-x^2\right)^{1/2}(x^2-t^2)} dx \right\} \right.$$

$$\left. + 2 \int_0^1 \left(\frac{x^2 f'(x)}{(1-x^2)^{1/2}} dx \right) \right], \quad 0 < t < 1 \tag{4.1.43}$$

after using the identity

$$x^2(1-x^2)^{1/2} = \frac{x^2(1-x^2)}{\left(1-x^2\right)^{1/2}} = \frac{(x^2-t^2+t^2)(1-t^2+t^2-x^2)}{\left(1-x^2\right)^{1/2}} \tag{4.1.44}$$

From the result (4.1.43), we thus observe that the singular integral equation (4.1.33) can have the solution, which is bounded at both the end points $t = 0$ and $t = 1$, if *and only if* the following two conditions are satisfied by the forcing term $f(x)$:

(i)

$$\int_0^1 \frac{f'(x)}{\left(1-x^2\right)^{1/2}} dx = 0 \tag{4.1.45}$$

and

(ii) $$c + 2 \int_0^1 \frac{x f'(x)}{\left(1 - x^2\right)^{1/2}} \, dx = 0$$

which, after using the relation (4.1.42), becomes

$$\int_0^1 \frac{x f(x) + x^2 f'(x)}{\left(1 - x^2\right)^{1/2}} \, dx = 0. \tag{4.1.46}$$

Thus, this bounded solution is given by

$$\varphi(t) = -\frac{2}{\pi} t (1 - t^2)^{1/2} \int_0^1 \frac{f'(x)}{\left(1 - x^2\right)^{1/2} \left(x^2 - t^2\right)} \, dx, \quad 0 < t < 1. \tag{4.1.47}$$

It is easily verified by using the given integral equation (4.1.33) and the equation (4.1.34) that the second condition (4.1.46) is satisfied *automatically*.

We can determine the two other forms of the solution $\varphi(t)$ of the singular integral equation (4.1.33), which are (*i*) bounded at the end point $t = 0$, but unbounded at $t = 1$, (*ii*) bounded at the end pointy t = 1, but unbounded at t = 0, by formally rewriting the general solution (4.1.38) in two other different forms, as was done in the relation (4.1.43), by using the identities

(i) $x^2 (1 - x^2)^{1/2} = (x^2 - t^2 + t^2)(1 - x^2)^{1/2}$

(ii) $x^2 (1 - x^2)^{1/2} = \dfrac{x^2 (1 - t^2 + t^2 - x^2)}{\left(1 - x^2\right)^{1/2}}$ \hfill (4.1.48)

respectively. We find by using the identity (i), that the solution (4.1.38) can be rewritten as

$$\varphi(t) = \frac{1}{\pi} \frac{1}{t \left(1 - t^2\right)^{1/2}} \left[c - 2 t^2 \int_0^1 \frac{\left(1 - x^2\right)^{1/2} f'(x)}{\left(x^2 - t^2\right)} \, dx \right.$$

$$\left. - 2 \int_0^1 (1 - x^2)^{1/2} f'(x) \, dx \right], \quad 0 < t < 1. \tag{4.1.49}$$

whereas, by using the identity (ii), the solution (4.1.38) can also be rewritten as

$$\varphi(t) = \frac{1}{\pi} \frac{1}{t\left(1-t^2\right)^{1/2}} \left[c - 2(1-t^2) \int_0^1 \frac{x^2 f'(x)}{\left(1-x^2\right)^{1/2}\left(x^2 - t^2\right)} dx \right.$$

$$\left. + 2 \int_0^1 \frac{x^2 f'(x)}{\left(1-x^2\right)^{1/2}} dx \right], \quad 0 < t < 1. \tag{4.1.50}$$

Thus, the two appropriate forms of the solutions for the particular cases (*i*) and (*ii*), mentioned above, are given by

(i) $\quad \varphi(t) = -\dfrac{2t}{\pi\left(1-t^2\right)^{1/2}} \displaystyle\int_0^1 \dfrac{\left(1-x^2\right)^{1/2} f'(x)}{x^2 - t^2} dx, \quad 0 < t < 1,$ (4.1.51)

provided that

$$c - 2 \int_0^1 \left(1-x^2\right)^{1/2} f'(x)\, dx = 0,$$

which produces, by using (4.1.42),

$$\int_0^1 \frac{(1-x^2) f'(x) - x f(x)}{\left(1-x^2\right)^{1/2}} dx = 0 \tag{4.1.52}$$

which can be shown to be satisfied *identically*, if use is made of the given integral equation (4.1.33) and the equation (4.1.34),

(ii) $\quad \varphi(t) = -\dfrac{2\left(1-t^2\right)^{1/2}}{\pi t} \displaystyle\int_0^1 \dfrac{x^2 f'(x)}{\left(1-x^2\right)^{1/2}\left(x^2 - t^2\right)} dx, \quad 0 < t < 1,$ (4.1.53)

provided that

$$c + 2 \int_0^1 \frac{x^2 f'(x)}{\left(1-x^2\right)^{1/2}} dx = 0,$$

which produces, again by using (4.1.42),

$$\int_0^1 \frac{x^2 f'(x) + x f(x)}{\left(1-x^2\right)^{1/2}} dx = 0, \tag{4.1.54}$$

which can again be shown to be identically satisfied by using (4.1.33) and (4.1.34).

Remark

It is emphasized that the conditions (4.1.46), (4.1.52) and (4.1.54) are just three identities associated with the given integral equation (4.1.33), under the circumstances considered and thus, these are not any extra conditions to be satisfied by the forcing function $f(x)$. The major conclusion that can be drawn out of the above discussion, concerning the singular integral equation (4.1.33), is that the corresponding homogeneous integral equation does not possess any non-trivial solution at all, and that the solution of the *non-homogeneous* integral equation, therefore, does not contain any arbitrary constant, like what happens in the case of singular integral equation of the Cauchy type.

4.2 INTEGRAL EQUATIONS WITH CAUCHY TYPE KERNELS

A special elementary method

A special elementary method to solve the first kind singular integral equation with Cauchy type kernel is explained here. Let us consider the problem of solving the singular integral equation

$$\int_0^1 \frac{\varphi(t)}{t-x} \, dt = f(x), \quad 0 < x < 1 \tag{4.2.1}$$

where $\{x(1-x)\}^{1/2} \varphi(x)$ remains bounded near the end points $x = 0$ and $x = 1$.

We use the following transformations:

$$t = \cos^2 \theta, \ x = \cos^2 \alpha, \ \mu(\theta) = 2\varphi \, (\cos^2 \theta) \sin\theta \, \cos\theta, \ \lambda(\theta) = f(\cos^2 \theta) \tag{4.2.2}$$

so that $0 \le \theta \le \pi / 2, \ 0 \le \alpha \le \pi/2$. Then the equation (4.2.1) transform to

$$\int_0^{\pi/2} \frac{\mu(\theta)}{\cos^2 \theta - \cos^2 \alpha} \, d\theta = -\lambda(\alpha), \quad 0 < \alpha < \pi/2. \tag{4.2.3}$$

Now, because of the requirements that the function $\mu(\theta)$ is bounded at $\theta = 0$ and $\theta = \pi/2$, we assume that $\mu(\theta)$ can be expanded into the uniformly convergent trigonometric series as given by

$$\mu(\theta) = \frac{1}{2} a_0 + \sum_{n=1}^{\infty} a_{2n} \cos 2n\theta, \quad 0 < \theta < \pi/2, \tag{4.2.4}$$

where a_{2n} $(n = 0,1,...)$ are unknown constants.

Substituting the expression (4.2.4) into the left side of the equation (4.2.3), interchanging the orders of integration and summation, we obtain

$$-\lambda(\alpha) = \frac{1}{2} a_0 \int_0^{\pi/2} \frac{d\theta}{\cos^2\theta - \cos^2\alpha} + \sum_{n=1}^\infty a_{2n} \int_0^{\pi/2} \frac{\cos 2n\theta}{\cos^2\theta - \cos^2\alpha} d\theta, \ 0 < \theta < \pi/2. \quad (4.2.5)$$

We now evaluate the integrals in the relation (4.2.5) by re-expressing the integrals in terms of contour integrals, to give

$$I_n \equiv \int_0^{\pi/2} \frac{\cos 2n\theta}{\cos^2\theta - \cos^2\alpha} d\theta = 2 \int_0^{\pi/2} \frac{\cos 2n\theta}{\cos 2\theta - \cos 2\alpha} d\theta$$

$$\qquad (4.2.6)$$

$$= \frac{1}{2} \int_{-\pi}^\pi \frac{\cos n\varphi}{\cos\varphi \ \cos\beta} = -\frac{i}{2} \int_C \frac{z^{2n} + 1}{z^n (z - e^{i\beta})(z - e^{-i\beta})} dz$$

where $z = e^{i\varphi}$, $\beta = 2\alpha$, and C is the unit circle in the complex z-plane, with indentations at the points $z = e^{-i\beta}$ and $z = e^{i\beta}$. We then use the Cauchy's residue theorem and derive that

$$I_n = \frac{\pi \ \sin 2n\alpha}{\sin 2\alpha} \quad n = 0,1,2,... \ . \quad (4.2.7)$$

We thus obtain, by using the results (4.2.7) in (4.2.5), that

$$-\lambda(\alpha) = \frac{\pi}{\sin 2\alpha} \sum_{n=1}^\infty a_{2n} \sin 2n\alpha, \ 0 < \alpha < \pi/2. \quad (4.2.8)$$

The coefficients a_{2n} $(n=1,2,...)$ in the series (4.2.8), are obtained, by utilizing the orthogonality of the functions $\sin 2n\alpha$ $(n = 1,2,...)$, as

$$a_{2n} = -\frac{8}{\pi^2} \int_0^{\pi/2} \lambda(\alpha) \sin 2\alpha \ \sin 2n\alpha \ d\alpha, \ n = 1,2,.... \quad (4.2.9)$$

Next, utilizing the relations (4.2.9) into (4.2.4), we obtain

$$\mu(\theta) = \frac{1}{2} a_0 - \frac{8}{\pi^2} \int_0^{\pi/2} \lambda(\alpha) \ \sin 2\alpha \ T(\alpha,\theta) \ d\theta \quad (4.2.10)$$

where

$$T(\alpha,\theta) = \lim_{x \to 1-0} \sum_{n=1}^{\infty} x^{2n} \sin 2n\alpha \, \cos 2n\theta$$

(4.2.11)

$$= \frac{\sin 2\alpha}{8(\cos^2\theta - \cos^2\alpha)}$$

obtained by writing

$$\sin 2n\alpha \, \cos 2n\theta = \frac{1}{2}\{\sin 2n(\alpha+\theta) + \sin 2n(\alpha-\theta)\}$$

and summing up the series by standard tricks. We thus determine $\mu(\theta)$ as given by

$$\mu(\theta) = \frac{1}{2}a_0 - \frac{4}{\pi^2}\int_0^{\pi/2} \frac{f(\cos^2\alpha) \, \cos^2\alpha \, (1-\cos^2\alpha)}{\cos^2\theta - \cos^2\alpha} \, d\alpha, \quad 0 < \theta < \pi/2. \quad (4.2.12)$$

Reverting back to the old variables (see the relations (4.2.2)), we obtain the complete solution of the given integral equation (4.2.1) as

$$\varphi(x) = \frac{1}{\{x(1-x)\}^{1/2}}\left[C_0 - \frac{1}{\pi^2}\int_0^1 \frac{\{t(1-t)\}^{1/2} f(t)}{t-x} \, dt\right], \quad 0 < x < 1, \quad (4.2.13)$$

where C_0 is an arbitrary constant, which can be identified as

$$C_0 = \frac{1}{\pi}\int_0^1 \varphi(x) \, dx. \quad (4.2.14)$$

This result agrees with the results obtained in earlier chapters.

We mention here that there has been a Lemma, proved by Ursell (1947), which is as follows:

Lemma   The singular integral equation

$$\int_a^{\infty} \frac{\mu(u)}{y^2 - u^2} \, du = \lambda(y), \quad a < y < \infty, \quad (4.2.15)$$

has, for suitable $\lambda(y)$, a solution $\mu(u)$ such that $(u^2 - a^2)^{1/2}\mu(u)$ is bounded near $u = a$, this solution being given by

$$\mu(u) = \frac{u}{(u^2 - a^2)^{1/2}}\left[C + \frac{4u^2}{\pi^2}\int_a^{\infty} \frac{(y^2-a^2)^{1/2}\lambda(y)}{y(y^2-u^2)} \, dy\right], \quad a < u < \infty, \quad (4.2.16)$$

where C is an arbitrary constant.

It may be verified that the transformations

$$u = a \; \sec\theta, \quad y = a \; \sec\alpha, \quad 0 \le \theta, \; \alpha \le \pi/2 \qquad (4.2.17)$$

cast the integral equation (4.2.15) into the one which is similar to the equation (4.2.3) and hence the above Lemma can be proved easily.

4.3 USE OF POINCARE'-BERTRAND FORMULA

Here we explain the use of the well-known Poincare'-Bertrand formula, involving repeated singular integrals of the Cauchy type, to derive the solutions of some important singular integral equations with Cauchy type singularities. This has already been introduced in the last chapter. A slightly different but equivalent form is given here.

Let L be a smooth arc or contour, $\varphi(t, t_1)$ a function of two points t and t_1 on L satisfying the Hölder condition with respect to t and t_1, and let t_0 be a fixed point on L not coinciding with one of the end points of L. Then the following formula, known as the Poincare'-Bertrand formula (PBF) holds good:

$$\int_L \frac{1}{t-t_0} \left\{ \int_L \frac{\varphi(t,t_1)}{t_1-t} \, dt_1 \right\} dt = -\pi^2 \varphi(t_0,t_0) + \int_L \left\{ \int_L \frac{\varphi(t,t_1)}{(t-t_0)(t_1-t)} \, dt \right\} dt_1. \quad (4.3.1)$$

As a *first use* of the PBE, we consider the problem of solving the following singular integral equation of the Cauchy type (cf. Chakrabarti (1980)).

Problem 1 Solve the singular integral equation

$$(T\varphi)\,(x) \equiv \int_{-1}^{1} \frac{\varphi(t)}{t-x} \, dt = f(x), \quad -1 < x < 1, \qquad (4.3.2)$$

where the functions $\varphi(x)$ and $f(x)$ are assumed to satisfy the Hölder conditions in $(-1,1)$.

Solution

We rewrite both sides of the equation (4.3.2) as

$$(1-x^2)^{1/2} \, (T\varphi)\,(x) = \left(1-x^2\right)^{1/2} f(x), \quad -1 < x < 1, \quad (4.3.3)$$

and then multiply both sides of the equation (4.3.3) by $\left(\dfrac{1-x}{1+x}\right)^{\frac{1}{2}-\beta} \dfrac{1}{x-\xi}$ for $-1<\xi<1$, and integrate with respect to x between $x=-1$ to $x=1$, where β is an unknown constant to be determined, as described below, such that $0 < Re\beta < 1$. Then interchanging the order of the repeated integrals, in the above procedure, by using the PBF (4.3.1), with $L=(-1,1)$, we arrive at the following result:

$$-(1-\xi)\left(\frac{1+\xi}{1-\xi}\right)^{\beta} \pi^2\varphi(\xi)+\int_{-1}^{1} \frac{\varphi(t)}{t-\xi}\left\{\int_{-1}^{1}(1-x)\left(\frac{1+x}{1-x}\right)^{\beta}\left(\frac{1}{t-x}+\frac{1}{x-\xi}\right)dx\right\}dt$$

$$= \int_{-1}^{1}\left(\frac{1-x}{1+x}\right)^{\frac{1}{2}-\beta} \frac{\left(1-x^2\right)^{1/2} f(x)}{x-\xi} dx, \quad -1<\xi<1.$$
(4.3.4)

If we now use the following result (cf. Gakhov (1966))

$$\int_{-1}^{1}(1-x)\left(\frac{1+x}{1-x}\right)^{\beta} \frac{dx}{x-\lambda} = (1-\lambda)\left[\frac{\pi}{\sin\pi\beta}-\pi\cot\pi\beta\left(\frac{1+\lambda}{1-\lambda}\right)^{\beta}\right]-C(\beta), \quad -1<\lambda<1 \quad (4.3.5)$$

where

$$C(\beta) = \int_{-1}^{1}\left(\frac{1+x}{1-x}\right)^{\beta} dx,$$
(4.3.6)

the relation (4.3.4) takes up the form

$$-(1-\xi)\left(\frac{1+\xi}{1-\xi}\right)^{\beta} \pi^2\varphi(\xi)+\frac{\pi}{\sin\pi\beta} \int_{-1}^{1} \varphi(t) dt +\pi\cot\pi\beta \int_{-1}^{1} \frac{\varphi(t)}{t-\xi}\left\{(1-t)\left(\frac{1+t}{1-t}\right)^{\beta}\right.$$

$$\left. -(1-\xi)\left(\frac{1+\xi}{1-\xi}\right)^{\beta}\right\}dt = \int_{-1}^{1}\left(\frac{1-x}{1+x}\right)^{\frac{1}{2}-\beta} \frac{\left(1-x^2\right)^{1/2} f(x)}{x-\xi}dx, \quad -1<\xi<1.$$
(4.3.7)

We thus find that the relation (4.3.7) is satisfied by the function $\varphi(x)$ of the given integral equation (4.3.2) for all constants β such that $0 < Re\beta <1$. It is this important relation (4.3.7), which will produce solution of the integral equation (4.3.2) by choosing the unknown constant β appropriately. We observe that the only choice for β is $\beta = \dfrac{1}{2}$, and then (4.3.7) produces

$$\varphi(\xi) = \frac{1}{\pi\left(1-\xi^2\right)^{1/2}}\left[A-\frac{1}{\pi} \int_{-1}^{1} \frac{\left(1-x^2\right)^{1/2} f(x)}{x-\xi}dx\right], \quad -1<\xi<1, \quad (4.3.8)$$

where

$$A = \int_{-1}^{1} \varphi(t)\, dt \qquad (4.3.9)$$

and can be regarded as an arbitrary constant since

$$\int_{-1}^{1} \frac{d\xi}{\left(1-\xi^2\right)^{1/2}(x-\xi)} = 0 \text{ for } -1<x<1.$$

It may be noted that if $\varphi(t)$ is an odd function, then $A = 0$.

As a second use of the PBF, we consider the following problem, occurring in the theory of dislocations (cf. Williams (1978)).

Problem 2 Solve the singular integral equation

$$\int_{0}^{1} \left(\frac{1}{\eta-\sigma} + \frac{p}{\eta+\sigma} \right) \psi(\eta)\, d\eta = h(\sigma), \quad 0<\sigma<1 \quad (4.3.10)$$

where $p\,(>-1)$ is a known constant, and ψ and h satisfy Hölder conditions in $(0,1)$.

Solution

We first use the following transformations:

$$\eta = \left(1-t^2\right)^{1/2}, \quad \sigma = \left(1-x^2\right)^{1/2} \text{ and } \varphi(t) = \left(1-t^2\right)^{1/2} \psi\left((1-t^2)^{1/2}\right),$$

$$(4.3.11)$$

$$g(x) = h\left((1-x^2)^{1/2}\right), \quad \varphi(-t) = -\varphi(t), \quad g(-t) = -g(t), \quad t > 0.$$

Then the given equation (4.3.10) gets transformed into the new singular integral equation, as given by

$$a^2 \left(1-x^2\right)^{1/2} \int_{-1}^{1} \frac{\varphi(t)}{t-x}\, dt + \int_{-1}^{1} \frac{\left(1-t^2\right)^{1/2} \varphi(t)}{t-x}\, dt = \frac{2}{p+1} g(x), \quad -1<x<1 \quad (4.3.12)$$

where

$$a^2 = \frac{1-p}{1+p}. \qquad (4.3.13)$$

We next multiply both sides of equation (4.3.12) by $\left(\dfrac{1-x}{1+x}\right)^{\frac{1}{2}-\beta}\dfrac{1}{x-\xi}$ for $-1<\xi<1,$ and integrate with respect to x between $x=-1$ to $x=1$, where β is an unknown constant to be determined, as explained below, with $0<Re\,\beta<1$, so that the solution of the integral equation (4.3.12) is determined. We obtain, then, by using the above procedure and by applying the PBF (4.3.1) with $L=(-1,1)$,

$$-(1-\xi)\left(\frac{1+\xi}{1-\xi}\right)^{\beta}\left(1+a^2\right)\pi^2\varphi(\xi)+a^2\int_{-1}^{1}\frac{\varphi(t)}{t-\xi}\left\{\int_{-1}^{1}(1-x)\left(\frac{1+x}{1-x}\right)^{\beta}\left(\frac{1}{t-x}+\frac{1}{x-\xi}\right)dx\right\}$$

$$+\int_{-1}^{1}\frac{\left(1-t^2\right)^{1/2}\varphi(t)}{t-\xi}\left\{\int_{-1}^{1}\left(\frac{1-x}{1+x}\right)^{\frac{1}{2}-\beta}\left(\frac{1}{t-x}+\frac{1}{x-\xi}\right)dx\right\}dt=g_1(\xi),\quad -1<\xi<1 \qquad (4.3.14)$$

where

$$g_1(\xi)=\frac{2}{p+1}\int_{-1}^{1}\left(\frac{1-x}{1+x}\right)^{\frac{1}{2}-\beta}\frac{g(x)}{x-\xi}\,dx. \qquad (4.3.15)$$

Changing β to $1-\beta$ in the relation (4.3.14), we get

$$-(1-\xi)\left(\frac{1+\xi}{1-\xi}\right)^{1-\beta}\left(1+a^2\right)\pi^2\varphi(\xi)+a^2\int_{-1}^{1}\frac{\varphi(t)}{t-\xi}\left\{\int_{-1}^{1}(1-x)\left(\frac{1+x}{1-x}\right)^{1-\beta}\left(\frac{1}{t-x}+\frac{1}{x-\xi}\right)dx\right\}dt$$

$$+\int_{-1}^{1}\frac{\left(1-t^2\right)^{1/2}\varphi(t)}{t-\xi}\left\{\int_{-1}^{1}\left(\frac{1-x}{1+x}\right)^{\beta-\frac{1}{2}}\left(\frac{1}{t-x}+\frac{1}{x-\xi}\right)dx\right\}dt=g_2(\xi),\quad -1<\xi<1, \qquad (4.3.16)$$

where

$$g_2(\xi)=\frac{2}{p+1}\int_{-1}^{1}\left(\frac{1-x}{1+x}\right)^{\beta-\frac{1}{2}}\frac{g(x)}{x-\xi}\,dx. \qquad (4..3.17)$$

We now evaluate the inner integrals in the relations (4.3.14) and (4.3.16). By using the result (4.3.5), we obtain

$$-(1-\xi)\left(\frac{1+\xi}{1-\xi}\right)^{\beta}(1+a^2)\pi^2\varphi(\xi)+\frac{\pi a^2}{\sin\pi\beta}\int_{-1}^{1}\varphi(t)\,dt$$

$$-\pi a^2\,\cot\pi\beta\int_{-1}^{1}\frac{\varphi(t)}{t-\xi}\left\{(1-\xi)\left(\frac{1+\xi}{1-\xi}\right)^{\beta}-(1-t)\left(\frac{1+t}{1-t}\right)^{\beta}\right\}dt$$

$$+\pi\,\tan\pi\beta\int_{-1}^{1}\frac{\left(1-t^2\right)^{1/2}\varphi(t)}{t-\xi}\left\{\left(\frac{1+\xi}{1-\xi}\right)^{\beta-\frac{1}{2}}-\left(\frac{1+t}{1-t}\right)^{\beta-\frac{1}{2}}\right\}dt$$

$$= g_1(\xi),\quad -1<\xi<1,$$

(4.3.18)

and

$$-(1-\xi)\left(\frac{1+\xi}{1-\xi}\right)^{1-\beta}(1+a^2)\,\pi^2\varphi(\xi)+\frac{\pi a^2}{\sin\pi\beta}\int_{-1}^{1}\varphi(t)\,dt$$

$$+\pi a^2\,\cot\pi\beta\int_{-1}^{1}\frac{\varphi(t)}{t-\xi}\left\{(1-\xi)\left(\frac{1+\xi}{1-\xi}\right)^{1-\beta}-(1-t)\left(\frac{1+t}{1-t}\right)^{1-\beta}\right\}dt$$

$$-\pi\,\tan\pi\beta\int_{-1}^{1}\frac{\left(1-t^2\right)^{1/2}\varphi(t)}{t-\xi}\left\{\left(\frac{1+\xi}{1-\xi}\right)^{\frac{1}{2}-\beta}-\left(\frac{1+t}{1-t}\right)^{\frac{1}{2}-\beta}\right\}dt$$

$$= g_2(\xi),\quad -1<\xi<1,$$

(4.3.19)

Multiplying the relation (4.3.18) by $\left(\frac{1-\xi}{1+\xi}\right)^{\beta}$ and the relation (4.3.19) by $\left(\frac{1-\xi}{1+\xi}\right)^{1-\beta}$, and adding, we obtain

$$-2\pi^2(1+a^2)(1-\xi)\varphi(\xi)=\left(\frac{1-\xi}{1+\xi}\right)^{\beta}g_1(\xi)+\left(\frac{1-\xi}{1+\xi}\right)^{1-\beta}g_2(\xi)$$

$$-\frac{\pi a^2 C}{\sin\pi\beta}\left\{\left(\frac{1-\xi}{1+\xi}\right)^{\beta}+\left(\frac{1-\xi}{1+\xi}\right)^{1-\beta}\right\}+\pi a^2\cot\pi\beta\int_{-1}^{1}\frac{\varphi(t)}{t-\xi}(1-t)\left\{\left(\frac{1-\xi}{1+\xi}\frac{1+t}{1-t}\right)^{\beta}\right.$$

$$\left.-\left(\frac{1-\xi}{1+\xi}\frac{1+t}{1-t}\right)^{1-\beta}\right\}dt-\pi\,\tan\pi\beta\int_{-1}^{1}\frac{\varphi(t)}{t-\xi}\left(1-t^2\right)^{1/2}\left\{\left(\frac{1-\xi}{1+\xi}\right)^{\beta}\left(\frac{1+t}{1-t}\right)^{\beta-\frac{1}{2}}\right.$$

$$\left.-\left(\frac{1-\xi}{1+\xi}\right)^{1-\beta}\left(\frac{1+t}{1-t}\right)^{\frac{1}{2}-\beta}\right\}dt,\quad -1<\xi<1,$$

(4.3.20)

where

$$C = \int_{-1}^{1}\varphi(t)\,dt.$$

(4.3.21)

Now, noting that

$$\left(1-t^2\right)^{1/2} = (1-t)\left(\frac{1+t}{1-t}\right)^{1/2},$$

we observe from the relation (4.3.20) that the solution φ of the integral equation (4.3.12) will be possible to be determined from the relation (4.3.20) itself, if we choose the constant β such that

$$a^2 \cot\pi\beta = \tan\pi\beta$$

which is equivalent to

$$\tan\pi\beta = |a|, \tag{4.3.22}$$

and, in that event, the solution to the equation (4.3.12) is

$$\varphi(t) = -\frac{1}{2\pi^2(1+a^2)(1-t)}\left[\left(\frac{1-t}{1+t}\right)^\beta g_1(t) + \left(\frac{1-t}{1+t}\right)^{1-\beta} g_2(t)\right.$$
$$\left. -\frac{\pi a^2 C}{\sin\pi\beta}\left\{\left(\frac{1-t}{1+t}\right)^\beta + \left(\frac{1-t}{1+t}\right)^{1-\beta}\right\}\right], \quad -1 < t < 1. \tag{4.3.23}$$

Now, in our given integral equation (4.3.12), φ is an odd function (cf. the transformation (4.3.11), and hence we have $C = 0$, so that the solution (4.3.23) finally takes up the form

$$\varphi(t) = -\frac{1}{2\pi^2(1+a^2)(1-t)}\left[\left(\frac{1-t}{1+t}\right)^\beta g_1(t) + \left(\frac{1-t}{1+t}\right)^{1-\beta} g_2(t)\right], \quad -1 < t < 1 \tag{4.3.24}$$

where the functions g_1 and g_2 are defined by the relations (4.3.15) and (4.3.17) respectively.

The form (4.3.24) of the solution of the integral equation (4.3.12) agrees with that obtained by Lewin (1961) by employing a different method.

4.4 SOLUTION OF SINGULAR INTEGRAL EQUATION INVOLVING TWO INTERVALS

Using the well known inverse of the Cauchy singular operator on a single interval, derived in section 3, along with certain standard results associated with singular integrals, the general solution of a singular integral equation in a double interval, whose kernel involves a combination of logarithmic and Cauchy type singularities, is obtained here.

We consider the singular integral equation

$$\int_L \varphi(t) \left\{ K \ln \left| \frac{t-x}{t+x} \right| + \frac{1}{x+t} + \frac{1}{x-t} \right\} dt = f(x), \quad x \in L, \tag{4.4.1}$$

where $K > 0$, $f(x)$ is a prescribed function, L is the union of two intervals $(0, a)$ and (b, ∞) $(a < b)$, and

$$\varphi(t) = 0 \left(|t - c| \right)^{1/2} \text{ as } t \to c \text{ where } c = a, b. \tag{4.4.2}$$

This integral equation arises in the study of surface water wave scattering by a fully immersed thin vertical plate of finite width (see Chakrabarti (1989) for similar studies). Banerjea and Mandal (1993) used a method to solve this integral equation which is somewhat involved.

Here we employ a method of some *special* type that avoids many details involving complex variable theory and the Riemann-Hilbert problem used in Chapter 3 for tackling such singular integral equations.

To solve the integral equation (4.4.1), we rewrite it in the following form

$$\int_0^a \varphi_1(t) \left\{ K \ln \left| \frac{t-x}{t+x} \right| + \frac{1}{x+t} + \frac{1}{x-t} \right\} dt + \int_b^\infty \varphi_2(t) \left\{ K \ln \left| \frac{t-x}{t+x} \right| + \frac{1}{x+t} + \frac{1}{x-t} \right\} dt \tag{4.4.3}$$

$$= f(x), \quad x \in L$$

where

$$\varphi(t) = \begin{cases} \varphi_1(t) & \text{for } 0 < t < a, \\ \varphi_2(t) & \text{for } b < t < \infty, \end{cases} \tag{4.4.4}$$

and

$$f(x) = \begin{cases} f_1(x) & \text{for } 0 < x < a, \\ f_2(x) & \text{for } b < x < \infty. \end{cases} \tag{4.4.5}$$

After making the substitutions

$$\psi_1(t) = -K \int_t^a \varphi_1(s) \, ds + \varphi_1(t), \quad 0 < t < a, \tag{4.4.6}$$

and

$$\psi_2(t) = K \int_b^t \varphi_2(s) \, ds + \varphi_2(t), \quad b < t < \infty, \tag{4.4.7}$$

the equation (4.4.3) can be expressed as

$$\int_0^a \frac{\Psi_1(t)}{x^2 - t^2} \, dt + \int_b^\infty \frac{\Psi_2(t)}{x^2 - t^2} \, dt = \frac{f(x)}{2x}, \quad x \in L. \qquad (4.4.8)$$

Letting

$$x^2 = \xi, \quad t^2 = \eta, \quad a^2 = \alpha, \quad b^2 = \beta,$$

$$-\frac{\Psi_1(t)}{t} = \lambda_1(\eta), \quad -\frac{\Psi_2(t)}{t} = \lambda_2(\eta), \quad \text{and} \quad \frac{f(x)}{x} = g(\xi), \qquad (4.4.9)$$

the equation (4.4.8) can be rewritten as

$$\int_0^\alpha \frac{\lambda_1(\eta)}{\eta - \xi} \, d\eta + \int_\beta^\infty \frac{\lambda_2(\eta)}{\eta - \xi} \, d\eta = g(\xi), \quad \xi \in L' \qquad (4.4.10)$$

where L' is the union of the intervals $(0, \alpha)$ and (β, ∞). To solve the equation (4.4.10) we require the following definitions and results.

Definitions

Operators:

We define the operators $T_1, \tilde{T}_1, T_2$ and $\tilde{T}_2$ as follows:

(i) $$(T_1 f)(\xi) = \int_0^\alpha \frac{f(\eta)}{\eta - \xi} \, d\eta \quad \text{for } \xi \in (0, \alpha),$$

(ii) $$(\tilde{T}_1 f)(\xi) = \int_0^\alpha \frac{f(\eta)}{\eta - \xi} \, d\eta \quad \text{for } \xi \notin (0, \alpha),$$

(iii) $$(T_2 f)(\xi) = \int_\beta^\infty \frac{f(\eta)}{\eta - \xi} \, d\eta \quad \text{for } \xi \in (\beta, \infty),$$

(iv) $$(\tilde{T}_2 f)(\xi) = \int_\beta^\infty \frac{f(\eta)}{\eta - \xi} \, d\eta \quad \text{for } \xi \notin (\beta, \infty),$$

For simplicity, we use the following notations

(i) $\Delta_1(\xi) = \left(\dfrac{\xi}{\alpha - \xi}\right)^{1/2}$ for $0 < \xi < \alpha$, $\tilde{\Delta}_1(\xi) = \left(\dfrac{\xi}{\xi - \alpha}\right)^{1/2}$ for $\xi > \beta$

(ii) $\Delta_2(\xi) = (\xi - \beta)^{1/2}$ for $\xi > \beta$, $\tilde{\Delta}_2(\xi) = (\beta - \xi)^{1/2}$ for $0 < \xi < \beta$,

(iii) $\Delta_3(\xi) = \{\xi(\alpha - \xi)\}^{1/2}$ for $0 < \xi < \alpha$, $\tilde{\Delta}_3(\xi) = \{\xi(\xi - \alpha)\}^{1/2}$ for $\xi > \beta$.

The inverse operators:

The inverse operator T_1^{-1} is obtained as (see equation (3.4.28))

$$\left(T_1^{-1}h\right)(\xi) = \frac{1}{\Delta_3(\xi)}\left\{C_0 - \frac{1}{\pi^2}\,T_1(\Delta_3 h)(\xi)\right\}, \quad \xi \in (0,\alpha)$$

where C_0 is an arbitrary constant. Since

$$\frac{1}{\Delta_1(\xi)}T_1(\Delta_1 h)(\xi) = \frac{1}{\Delta_3(\xi)}\left\{T_1(\Delta_3 h)(\xi) + \text{constant}\right\}$$

we can write

$$\left(T_1^{-1}h\right)(\xi) = \frac{C_1}{\Delta_3(\xi)} - \frac{1}{\pi^2 \Delta_1(\xi)}T_1(\Delta_1 h)(\xi), \quad \xi \in (0,\alpha) \qquad (4.4.11)$$

where C_1 is also an arbitrary constant. We can use (4.4.11) as the definition of the inverse operator T_1^{-1}. The inverse operator T_2^{-1} can similarly be defined as

$$\left(T_2^{-1}h\right)(\xi) = \frac{1}{\Delta_2(\xi)}\left\{C_2 - \frac{1}{\pi^2}T_2(\Delta_2 h)(\xi)\right\}, \quad \xi \in (\beta,\infty), \qquad (4.4.12)$$

where C_2 is an arbitrary constant.

Results The following results can be proved easily (left as simple exercises):

1. $\left[T_1^{-1}(T_1 f)\right](\xi) = \dfrac{C_1}{\Delta_3(\xi)} + f(\xi)$

2.　$\left[T_2^{-1}\left(T_2 f\right)\right](\xi) = \dfrac{C_2}{\Delta_2(\xi)} + f(\xi).$

3.　$\left[T_1^{-1}\left(\tilde{T}_2 f\right)\right](\xi) = \dfrac{C_1}{\Delta_3(\xi)} - \dfrac{1}{\pi\Delta_1(\xi)}\left[\tilde{T}_2\left(\tilde{\Delta}_1 f\right)\right](\xi).$

4.　$\left[T_2^{-1}\left(\tilde{T}_1 f\right)\right](\xi) = \dfrac{C_2}{\Delta_2(\xi)} - \dfrac{1}{\pi\Delta_2(\xi)}\left[\tilde{T}_1\left(\tilde{\Delta}_2 f\right)\right](\xi).$

5.　$\left[\tilde{T}_j\left\{\dfrac{1}{\Delta_j}\tilde{T}_k\left(\tilde{\Delta}_j f\right)\right\}\right](\xi) = \dfrac{\pi}{\tilde{\Delta}_j(\xi)}\left[T_k\left(\tilde{\Delta}_j f\right)\right](\xi) - \pi\,(T_k f)(\xi),$

　　where $j \neq k,\ j = 1, 2$ and $k = 1, 2.$

6.　$\left[\tilde{T}_j\left\{\dfrac{1}{\Delta_j}T_j\left(\Delta_j f\right)\right\}\right](\xi) = \dfrac{\pi}{\tilde{\Delta}_j(\xi)}\left[\tilde{T}_j\left(\Delta_j f\right)\right](\xi),\ j = 1, 2.$

7.　$\left[T_1\left(\dfrac{1}{\Delta_1}\right)\right](\xi) = -\pi.$

8.　$\left[\tilde{T}_1\left(\dfrac{1}{\Delta_1}\right)\right](\xi) = \pi\left\{\dfrac{1}{\tilde{\Delta}_1(\xi)} - 1\right\}.$

9.　$\left[\tilde{T}_1\left(\dfrac{1}{\Delta_3}\right)\right](\xi) = -\dfrac{\pi}{\tilde{\Delta}_3(\xi)}.$

10.　$\left[\tilde{T}_2\left(\dfrac{1}{\Delta_2}\right)\right](\xi) = \dfrac{\pi}{\tilde{\Delta}_2(\xi)}.$

C_1 and C_2 are arbitrary constants in all the above results.

The equation (4.4.10) can be written as a system of equations as given by

$$(T_1\lambda_1)(\xi) + \left(\tilde{T}_2\lambda_2\right)(\xi) = h_1(\xi),\ \ 0 < \xi < \alpha \qquad (4.4.13)$$

and

$$\left(\tilde{T}_1\lambda_1\right)(\xi) + (T_2\lambda_2)(\xi) = h_2(\xi),\ \ \xi > \beta \qquad (4.4.14)$$

where

$$h_1(\xi) = \frac{g_1(\xi^{1/2})}{\xi^{1/2}}, \quad h_2(\xi) = \frac{g_2(\xi^{1/2})}{\xi^{1/2}}. \tag{4.4.15}$$

Applying the operator T_2^{-1}, as defined by the relation (4.4.12), on both sides of the equation (4.4.14) and using the results 2 and 4 above, we get

$$\lambda_2(\xi) = \frac{1}{\Delta_2(\xi)} \left[\frac{1}{\pi} \{\tilde{T}_1(\tilde{\Delta}_2 \lambda_1)\}(\xi) - \frac{1}{\pi^2} \{T_2(\Delta_2 h_2)\}(\xi) - C_2 \right]. \tag{4.4.16}$$

Using the expression (4.4.16) in the equation (4.4.13) along with the results 5, 6 and 10 above, we get

$$\left[T_1(\tilde{\Delta}_2 \lambda_1) \right](\xi) = \frac{1}{\pi} \{\tilde{T}_2(\Delta_2 h_2)\}(\xi) + \pi C_2 + \tilde{\Delta}_2(\xi) h_1(\xi). \tag{4.4.17}$$

Again, applying the operator T_1^{-1}, on both sides of the equation (4.4.17), along with the results 1, 3 and 11 above, we get

$$\lambda_1(\xi)\tilde{\Delta}_2(\xi) = \frac{(1+\pi)C_1}{\pi \, \Delta_3(\xi)} - \frac{1}{\Delta_1(\xi)} \left[C_2 + \frac{1}{\pi^2} [T_1(\Delta_1 \tilde{\Delta}_2 h_1)](\xi) + [\tilde{T}_2(\Delta_2 \tilde{\Delta}_1 h_2)](\xi) \right]. \tag{4.4.18}$$

After a little simplification, the equation (4.4.18) can be expressed as

$$\lambda_1(\xi) = -\frac{A_1\xi + B_1}{\pi \{\xi(\alpha-\xi)(\beta-\xi)\}^{1/2}} - \frac{1}{\pi^2\xi^{1/2}} \left(\frac{\alpha-\xi}{\beta-\xi}\right)^{1/2} \left[\left[T_1\left(\frac{\eta^{1/2}(\beta-\eta)^{1/2}}{(\alpha-\eta)^{1/2}} h_1(\eta)\right) \right](\xi) \right. $$
$$\left. + \left[\tilde{T}_2\left(\frac{\eta^{1/2}(\eta-\beta)^{1/2}}{(\eta-\alpha)^{1/2}} h_2(\eta)\right) \right](\xi) \right], \quad 0 < \xi < \alpha, \tag{4.4.19}$$

where

$$A_1 = -\pi C_2, \quad B_1 = \pi \alpha C_2 - (1+\pi)C_1. \tag{4.4.20}$$

and these represent two arbitrary constants. Using the expression on the right side of the equation (4.4.18), in the equation (4.3.16) and utilizing the results 5, 6, 8 and 9, we arrive at the expression for the unknown function $\lambda_2(\xi)$ as given by

$$\lambda_2(\xi) = -\frac{A_1\xi + B_1}{\pi \{\xi(\xi-\alpha)(\xi-\beta)\}^{1/2}} - \frac{1}{\pi^2\xi^{1/2}} \left(\frac{\xi-\alpha}{\xi-\beta}\right)^{1/2} \left[\left[\tilde{T}_1\left(\frac{\eta^{1/2}(\beta-\eta)^{1/2}}{(\alpha-\eta)^{1/2}} h_1(\eta)\right) \right](\xi) \right. $$
$$\left. + \left[T_2\left(\frac{\eta^{1/2}(\eta-\beta)^{1/2}}{(\eta-\alpha)^{1/2}} h_2(\eta)\right) \right](\xi) \right], \quad \xi > \beta. \tag{4.4.21}$$

Thus, the solution of the equation (4.4.10) is given by the expressions in the relations (4.4.19) and (4.4.21) for $0 < \xi < \alpha$ and $\xi > \beta$, respectively. Using the transformations as given by the relations (4.4.4) and (4.4.15) in the solutions (4.4.19) and (4.4.21), we finally obtain the solutions corresponding to the equation (4.4.8) as given by

$$\psi_1(x) = \frac{A_1 x^2 + B_1}{\pi \left\{ \left(a^2 - x^2\right)\left(b^2 - x^2\right)\right\}^{1/2}} + \frac{2}{\pi^2} \left(\frac{a^2 - x^2}{b^2 - x^2}\right)^{1/2} P(x), \quad 0 < x < a, \quad (4.4.22)$$

and

$$\psi_2(x) = -\frac{A_1 x^2 + B_1}{\pi \left\{ \left(x^2 - a^2\right)\left(x^2 - b^2\right)\right\}^{1/2}} + \frac{2}{\pi^2} \left(\frac{x^2 - a^2}{x^2 - b^2}\right)^{1/2} P(x), \quad x > b, \quad (4.4.23)$$

where

$$P(x) = \int_0^a \left(\frac{b^2 - t^2}{a^2 - t^2}\right)^{1/2} \frac{t f_1(t)}{t^2 - x^2} dt + \int_b^\infty \left(\frac{t^2 - b^2}{t^2 - a^2}\right)^{1/2} \frac{t f_2(t)}{t^2 - x^2} dt. \quad (4.4.24)$$

Using the expressions (4.4.6) and (4.4.7) along with the solutions (4.4.22) and (4.4.23), we obtain the explicit solution of the integral equation (4.4.1) as given by the formulae

$$\varphi(x) = \begin{cases} \varphi_1(x) = \dfrac{d}{dx}\left[e^{-Kx} \displaystyle\int_a^x e^{Ku}\, \psi_1(u)\, du \right], & 0 < x < a \\[4mm] \varphi_2(x) = \dfrac{d}{dx}\left[e^{-Kx} \displaystyle\int_a^x e^{Ku}\, \psi_2(u)\, du \right], & x > b. \end{cases} \quad (4.4.25)$$

This method of solution is given by Chakrabarti and George (1994). The results (4.4.25) fully agree with those obtained by Banerjea and Mandal (1993). In the case of nonhomogeneous equation (4.4.1), for which $f(x) \equiv 0$, the solutions are obvious from the results (4.4.22) and (4.4.23), and are in full agreement with the results obtained by Estrada and Kanwal (1987,1989).

Chapter 5

Hypersingular Integral Equations

In this chapter we explain the occurrence and usefulness of *hypersingular integral equations* in various branches of applied mathematics. Starting with the basic definition of hypersingular integrals, we will build the subject matter of this topic in order to apply it to certain specific problems involving cracks in an elastic medium and scattering of surface water waves by obstacles in the form of thin barriers.

5.1 DEFINITIONS

Starting with the singular integral of the Cauchy type, for a Hölder continuous function $\varphi(t)$, for $t \in (a,b)$, as defined by the relation

$$\int_a^b \frac{\varphi(t)}{t-x}\, dt = \lim_{\varepsilon \to +0}\left[\int_a^{x-\varepsilon} \frac{\varphi(t)}{(t-x)}\, dt + \int_{x+\varepsilon}^b \frac{\varphi(t)}{t-x}\, dt \right], \quad a<x<b, \quad (5.1.1)$$

we define the *hypersingular integral of order 2* for the same function $\varphi(t)$, as

$$\int_a^b \frac{\varphi(t)}{(t-x)^2}\, dt = \lim_{\varepsilon \to +0}\left[\int_a^{x-\varepsilon} \frac{\varphi(t)}{(t-x)^2}\, dt + \int_{x+\varepsilon}^b \frac{\varphi(t)}{(t-x)^2} + \frac{\varphi(x-\varepsilon)+\varphi(x+\varepsilon)}{\varepsilon} \right], \quad a<x<b. \quad (5.1.2)$$

In the circumstances when $\varphi(t)$ is also differentiable in (a, b), we can relate the two integrals in (5.1.1) and (5.1.2) as

$$\int_a^b \frac{\varphi(t)}{(t-x)^2}\, dt = \frac{d}{dx}\left[\int_a^b \frac{\varphi(t)}{t-x}\, dt \right], \quad a<x<b. \quad (5.1.3)$$

As an example, we find that for $\varphi(t)=1$, $a=-1$ and $b=1$ we obtain that

$$\int_{-1}^{1} \frac{1}{(t-x)^2} \, dt = \lim_{\varepsilon \to +0} \left\{ -\left[\frac{1}{t-x} \right]_{t=-1}^{t=x-\varepsilon} - \left[\frac{1}{t-x} \right]_{t=x+\varepsilon}^{t=1} - \frac{2}{\varepsilon} \right\} \quad (5.1.4)$$

$$= -\frac{2}{1-x^2}, \quad -1 < x < 1.$$

We then observe that, as a special case, we have

$$\int_{-1}^{1} \frac{1}{t^2} \, dt = -2 \quad (5.1.5)$$

showing that the value of the hypersingular integral $\int_{-1}^{1} \frac{1}{t^2} \, dt$, involving the positive function $\frac{1}{t^2}$ is a *negative number* (!), which is not an absurd result in this area of mathematics involving hypersingular integrals, the reason for which is attributed to the finite part of the otherwise divergent integral, as introduced by Hadamard (1952).

The *general* hypersingular integral of order $(n+1)$ $(n \geq 1)$ can be defined by means of the relation (cf. Fox (1957)).

$$\int_{a}^{b} \frac{\varphi(t)}{(t-x)^{n+1}} \, dt = \lim_{\varepsilon \to +0} \left[\int_{a}^{x-\varepsilon} \frac{\varphi(t)}{(t-x)^{n+1}} \, dt + \int_{x+\varepsilon}^{b} \frac{\varphi(t)}{(t-x)^{n+1}} \, dt - H_n(x,\varepsilon) \right], \quad a < x < b, \quad (5.1.6)$$

where

$$H_n(x,\varepsilon) = \sum_{k=0}^{n-1} \frac{1}{k!} \frac{\varphi^{(k)}(x-\varepsilon) - (-1)^{n-k} \varphi^{(k)}(x+\varepsilon)}{(n-k) \, \varepsilon^{n-k}} \quad (5.1.7)$$

whenever the kth derivative $\varphi^{(k)}(x)$ of $\varphi(x)$ is Hölder continuous in (a,b) for $k = 0,1,...n-1$.

The following theorem is extremely useful in dealing with hypersingular integrals (cf. Fox (1957)).

Theorem 5.1: If $\varphi \in C^{n,\mu}(a,b)$, i.e. $\varphi(t)$ possesses a Hölder continuous derivative of order n with

$$\left| \varphi^{(n)}(t_1) - \varphi^{(n)}(t_2) \right| < A \left| t_1 - t_2 \right|^{\mu}$$

whenever $t_1, t_2 \in (a,b)$, A is a positive constant and $0 < \mu < 1$, then the limit defined in the relation (5.1.6) exists, and we say that the hypersingular integral $\displaystyle\int_a^b \frac{\varphi(t)}{(t-x)^{n+1}} \, dt$ of order $(n+1)$ exists.

Proof: Using the property that $\varphi^{(n)}(t)$ exists for $t \in (a,b)$, we write by utilizing the Taylor's theorem, that for $x \in (a,b)$,

$$\varphi(t) = \varphi(x) + \frac{t-x}{1!}\varphi^{(1)}(x) + \frac{(t-x)^2}{2!}\varphi^{(2)}(x) + \cdots + \frac{(t-x)^{n-1}}{(n-1)!}\varphi^{(n-1)}(x) + \frac{(t-x)^n}{n!}\varphi^{(n)}(\theta) \quad (5.1.8)$$

whenever θ lies between t and x. The substitution of the relation (5.1.8), along with the relation (5.1.7) into the expression inside the square bracket in (5.1.6), is legitimate as it is clearly observed that the limit exists. This completes the proof of the theorem.

Remark

It is obvious that the definition (5.1.6) of hypersingular integrals of order $(n+1)$ can be extended to the case when the integration is taken along the arc of a plane curve and the variables involved are all complex, with a and b representing end points of the arc of integration, x being any fixed part on the arc.

Followings are some important properties of hypersingular integrals:

Property 1: If $\varphi \in C^{n,k}$ (a,b) (a,b being real numbers and $a<b$), and if $0 < m \le n$ ($n \ge 1$), then

$$\int_a^b \frac{\varphi(t)}{(t-x)^{n+1}} \, dt = \sum_{k=0}^{n-1} \frac{(n-k-1)!}{n!}\left[\frac{\varphi^{(k)}(a)}{(a-x)^{n-k}} - \frac{\varphi^{(k)}(b)}{(b-x)^{n-k}}\right]$$
$$+ \frac{(n-m)!}{n!}\int_a^b \frac{\varphi^{(m)}(t)}{(t-x)^{n-m+1}} \, dt, \quad a < x < b. \quad (5.1.9)$$

The extensions of the above property 1, to the case of general complex integration are the followings.

Property 1A: If $\varphi(z)$ is single valued and analytic in a domain which includes a simple closed Jordan curve C and its interior, then

$$\int_C \frac{\varphi(z)}{(z-u)^{n+1}} \, dz = \frac{(n-m)!}{n!}\int_C \frac{\varphi^m(z)}{(z-u)^{n-m+1}} \, dz, \quad m \le n \quad (5.1.10)$$

where the integrals are taken once round C and u is a fixed point on C.

Property 1B:

If (i) $\varphi \in C^{n,\mu}(-\infty, \infty)$ for some μ such that $0 < \mu < 1$, (ii) for large $|t|$, $\varphi^{(m)}(t) = 0\left(t^{n-m-p}\right)$, $(p > 0, \ m \leq n)$, and (iii) $\varphi^{(k)}(t)/t^{n-k} \to 0$ $(k = 0, 1..., m-1)$ as $t \to \pm\infty$, then

$$\int_{-\infty}^{\infty} \frac{\varphi(t)}{(t-x)^{n+1}} \, dt = \frac{(n-m)!}{n!} \int_{-\infty}^{\infty} \frac{\varphi^{(m)}(t)}{(t-x)^{n-m+1}} \, dt, \ m \leq n. \quad (5.1.11)$$

Property 2: If the conditions, involving the property $1B$, on the function $\varphi(t)$ are valid, and if for large $|t|$, $\varphi^{(n)}(t) = 0(t^{-1/2-p})$ for $p > 0$, then the following results, extending the Hilbert transform, hold good.

If $g(x) = \dfrac{1}{\pi} \displaystyle\int_{-\infty}^{\infty} \dfrac{\varphi(t)}{(t-x)^{n+1}} \, dt, \quad -\infty < x < \infty,$ then

i) $g(x) \in L^2(-\infty, \infty),$

ii) $\varphi^{(n)}(x) = -\dfrac{n!}{\pi} \displaystyle\int_{-\infty}^{\infty} \dfrac{g(t)}{t-x} \, dt$

and

iii) $\displaystyle\int_{-\infty}^{\infty} \{g(x)\}^2 \, dx = \dfrac{1}{(n!)^2} \displaystyle\int_{-\infty}^{\infty} \{\varphi^{(n)}(x)\}^2 \, dx.$

The extensions of the Plemelj's formulae for Cauchy integrals to hypersingular integrals are given by the following general property.

Property 3: If (i) for all points of a positively oriented arc C in the complex z-plane except possibly at the end points of C, $\varphi(z)$ possesses derivatives upto order n, (ii) $\varphi(z)$ satisfies Hölder condition on C and t is a point on C, other than the end points of C, then

$$\Phi^{\pm}(t) = \pm \frac{\varphi^{(n)}(t)}{2(n!)} + \frac{1}{2\pi i} \int_{C} \frac{\varphi(\tau)}{(\tau - t)^{n+1}} \, d\tau \quad (5.1.12)$$

where $\Phi^{\pm}(t)$ are the limiting values of the function

$$\Phi(z) = \frac{1}{2\pi i} \int_{C} \frac{\varphi(\tau)}{(\tau - z)^{n+1}} \, dt, \ z \notin C,$$

as z approaches C from the left or from the right of C.

5.2 OCCURRENCES OF HYPERSINGULAR INTEGRAL EQUATIONS

Many two-dimensional boundary value problems in mathematical physics can be formulated as hypersingular integral equations. In Chapter 1, how the problem of two-dimensional flow post a rigid plate in an infinite fluid gives rise to a simple hypersingular integral equation is demonstrated. Some further examples involving acoustic scattering by a hard plate, water wave interaction with thin impermeable barriers, stress fields around cracks, in which hypersingular integral equations occur, are now given

Problem 1: Potential flow past a flat plate

We consider the problem of an irrotational two-dimensional flow in an ideal fluid past a rigid flat plate L, say. Assuming that the total velocity potential can be expressed as

$$\varphi^{tot}(x, y) = \varphi_0(x, y) + \varphi(x, y) \tag{5.2.1}$$

where $\varphi_0(x, y)$ is the known velocity potential of the flow in the absence of the plate L and $\varphi(x, y)$ is the perturbed potential due to the presence of the plate, we can determine the potential function $\varphi(x, y)$, by using Green's identity, as described below.

We use rectangular co-ordinates (t, n) where t is along the plate L and n is normal to the plate, and let

$$\varphi(x, y) = \frac{\partial^2 \psi}{\partial n^2}(x, y) \tag{5.2.2}$$

where $\dfrac{\partial}{\partial n}$ denotes outward normal derivative on L, then ψ is the *reduced* potential, and satisfies the two-dimensional Laplace equation

$$\frac{\partial^2 \psi}{\partial x^2} + \frac{\partial^2 \psi}{\partial y^2} = 0 \tag{5.2.3}$$

in the fluid region. Note that φ also satisfies the Laplace equation. Thus $\dfrac{\partial \psi}{\partial n}$ also satisfies the Laplace equation in the fluid region. Hence the harmonic function $\dfrac{\partial \psi}{\partial n}$ at a point $p = (x, y)$ in the fluid region can be represented by using Green's integral theorem, in the form

$$\frac{\partial \psi}{\partial n}(x, y) = -\frac{1}{2\pi} \int_L \left[\frac{\partial^2 \psi(q)}{\partial n^2} \right] G(p, q) \, ds_q \tag{5.2.4}$$

where $q \equiv (\xi, \eta)$ is a point on the plate, and

$$G(p;q) = \frac{1}{2} \ln \left\{ (x-\xi)^2 + (y-\eta)^2 \right\}^{1/2} \tag{5.2.5}$$

is the Green's function satisfying the Laplace equation everywhere except at the point q, $\left[\dfrac{\partial^2 \psi(q)}{\partial n^2} \right]$ denoting the jump across L at the point q.

Differentiating (5.2.4) along the direction of n at (x, y) and using (5.2.2), we obtain

$$\varphi(x, y) = -\frac{1}{2\pi} \frac{\partial}{\partial n} \int_L [\varphi(q)] \, G(p,q) \, ds_q. \tag{5.2.6}$$

Using the condition of vanishing of the normal derivative of $\varphi^{tot}(x, y)$ on the rigid plate L, we obtain from (5.2.6) that

$$\frac{1}{2\pi} \frac{\partial^2}{\partial n^2} \int_L [\varphi(q)] \, G(p,q) \, ds_q = \frac{\partial \varphi_0}{\partial n} \, (p), \quad p \in L. \tag{5.2.7}$$

Now the derivative operation $\dfrac{\partial^2}{\partial n^2}$ can be interchanged with the integration operation provided the integral is interpreted as a hypersingular integral, and thus the equation (5.2.7) produces the hypersingular integral equation

$$\frac{1}{2\pi} \int_L [\varphi(q)] \frac{\partial^2 G}{\partial n^2} \, (p,q) \, ds_q = \frac{\partial \varphi_0}{\partial n} \, (p), \quad p \in L. \tag{5.2.8}$$

which is to be solved subject to the requirement that the potential difference $[\varphi(q)]$ across L vanishes at the two end points of L.

Remark

Since $\displaystyle\int_L [\varphi(q)] \, G(p,q) \, ds_q$ satisfies the Laplace equation

$\left(\dfrac{\partial^2}{\partial t^2} + \dfrac{\partial^2}{\partial n^2} \right)(\cdot) = 0$, the equation (5.2.7) can also be written as

$$\frac{1}{2\pi} \frac{\partial^2}{\partial t^2} \int_L [\varphi(q)] \, G(p,q) \, ds_q = -\frac{\partial \varphi_0}{\partial n} \, (p), \quad p \in L. \tag{5.2.9}$$

where $\dfrac{\partial}{\partial t}$ denotes the tangential derivative. We note that the equation (5.2.9), which is equivalent to the equation (5.2.7) giving rise to the hypersingular integral equation (5.2.8), can be integrated out along the line L, and thus we can derive a *weakly singular* integral equation, as given by

$$\frac{1}{2\pi} \int_{L} [\varphi(q)]\, G(p,q)\, ds_q = f(p), \quad p \in L. \qquad (5.2.10)$$

where $f(p)$ involves two arbitrary constants of integration, A_1 and A_2 say. Thus, one can avoid bringing into picture hypersingular integral equation altogether provided that we can determine these two arbitrary constants completely. This approach, however, will not be pursued here.

Problem 2: Acoustic scattering by a hard plate

Let a plate occupying the position $y = 0$, $0 < x < a$, be immersed in an infinite compressible fluid. A time-harmonic incident sound wave described by the velocity potential $\mathrm{Re}\{\varphi_0(x,y)e^{-i\sigma t}\}$ be scattered by the plate. Henceforth the factor $e^{-i\sigma t}$ will be dropped throughout. Let $\varphi(x,y)$ denote the scattered potential. Then $\varphi(x,y)$ satisfies the differential equation,

$$\left(\frac{\partial^2}{\partial x^2} + \frac{\partial^2}{\partial y^2} + k^2\right)\varphi = 0 \quad \text{in the fluid region,}$$

the plate condition

$$\frac{\partial \varphi}{\partial y} = -\frac{\partial \varphi_0}{\partial y} \quad \text{on } y = 0, \ \ 0 < x < a,$$

the radiation condition,

$$\frac{\partial \varphi}{\partial r} - ik\varphi = 0 \quad \text{as } r \to \infty.$$

Let $G(x,y;\xi,\eta) = -\dfrac{1}{2}\, i\pi\, H_0^{(1)}(kR)$ with $R = \{(x-\xi)^2 + (y-\eta)^2\}^{1/2}$. Then $G \to \ln R$ as $R \to 0$ and satisfies the radiation condition at infinity.

By an appropriate use of Green's integral theorem to $\varphi(x,y)$ and $G(x,y;\xi,\eta)$ we obtain

$$\varphi(\xi,\eta) = -\frac{1}{2\pi} \int_0^a \psi(x) \frac{\partial G}{\partial y}(x,0;\xi,\eta) \, dx$$

where

$$\psi(x) = \varphi(x,+0) - \varphi(x,-0), \quad 0 < x < a,$$

and is unknown, and satisfies

$$\psi(0) = 0, \quad \psi(a) = 0.$$

We now use the condition on the plate

$$\frac{\partial \varphi}{\partial n}(\xi,0) = -\frac{\partial \varphi_0}{\partial y}(\xi,0), \quad 0 < \xi < a$$

in which the right side is a known function.

Now

$$\frac{\partial \varphi}{\partial n}(\xi,0) = \frac{1}{2\pi} \int_0^a \psi(x) \frac{i\pi k}{2} \frac{H_1^{(1)}(k|x-\xi|)}{|x-\xi|} \, dx$$

$$= \frac{1}{2\pi} \int_0^a \psi(x) \left[\frac{1}{(x-\xi)^2} + L(x,\xi) \right] dx$$

where

$$L(x,\xi) = \frac{i\pi k}{2} \frac{H_1^{(1)}(k|x-\xi|)}{|x-\xi|} - \frac{1}{(x-\xi)^2}$$

$$\sim -\frac{1}{2} k^2 \ln|x-\xi| \quad as \quad |x-\xi| \to 0.$$

Thus the equation for $\psi(x)$ is

$$\int_0^a \psi(x) \left[\frac{1}{(x-\xi)^2} + L(x,\xi) \right] dx = v(\xi), \quad 0 < \xi < a$$

where $v(\xi) = -2\pi \dfrac{\partial \varphi_0}{\partial n}(\xi,0)$ is a known function.

Problem 3: *Surface water wave scattering by a thin vertical barrier*

We consider here the two-dimensional problem of scattering of surface water waves described in Chapter 1. For the sake of completeness, the corresponding mathematical problem is described fully.

To solve the partial differential equation

$$\nabla^2 \varphi \equiv \frac{\partial^2 \varphi}{\partial x^2} + \frac{\partial^2 \varphi}{\partial y^2} = 0, \quad -\infty < x < \infty, \ y > 0 \qquad (5.2.11)$$

along with

i)   the free surface condition:

$$K\varphi + \frac{\partial \varphi}{\partial y} = 0 \ \text{ on } \ y = 0, \quad -\infty < x < \infty,$$

ii)   the condition on the barrier:

$$\frac{\partial \varphi}{\partial x} = 0 \ \text{ on } \ x = \pm 0, \ y \in L$$

where L is the portion of the y-axis, representing the barrier lying on the plane $x = 0$,

iii)   the condition on the gap:

φ and $\dfrac{\partial \varphi}{\partial x}$ are continuous across the gap $(x = 0, y \in \bar{L}$ where

$\bar{L} = (0, \infty) - L)$,

iv)   the surface wave conditions:

$$\varphi(x, y) \rightarrow \begin{cases} e^{-Ky+iKx} + R\, e^{-Ky-iKx} \\ T\, e^{-Ky+iKx} \ \text{ as } \ x \rightarrow \infty \end{cases}$$

where R and T are unknown constants, known as the complex reflection and transmission coefficients respectively,

v)   the condition at infinite depth:

$\varphi, \ \nabla \varphi \rightarrow 0 \ \text{ as } \ y \rightarrow \infty$

and

vi)   the edge condition(s):

$r^{1/2} \ \nabla \varphi = 0(1) \ \text{ as } \ r \rightarrow 0$

where r is the distance from a submerged edge of the barrier.

An appropriate solution of the problem satisfying the conditions (iv) and (v) and the continuity of $\dfrac{\partial \varphi}{\partial x}$ across $x = 0,\ y > 0$, is represented as

$$\varphi(x, y) = \begin{cases} (1-R)\, e^{-Ky+iKx} + \dfrac{\partial^2}{\partial y^2} \displaystyle\int_0^\infty \dfrac{A(k)}{k^2}\, e^{-kx}\, M(k,y)\, dk, & x > 0 \\[4mm] e^{-Ky+iKx} + R\, e^{-Ky-iKx} - \dfrac{\partial^2}{\partial y^2} \displaystyle\int_0^\infty \dfrac{A(k)}{k^2}\, e^{kx}\, M(k,y)\, dk, & x < 0 \end{cases} \tag{5.2.12}$$

where $A(k)$ is an unknown function, to be determined, and

$$M(k, y) = k \cos ky - K \sin ky. \tag{5.2.13}$$

It may be noted that the continuity of $\dfrac{\partial \varphi}{\partial x}$ across $x = 0,\ y > 0$, produces the equality

$$R + T = 1 \tag{5.2.14}$$

relating the unknown constant R and T, by using Havelock's inversion theorem (cf. Mandal and Chakrabarti (2000)).

If we now set

$$\psi(y) = \varphi(+0, y) - \varphi(-0, y), \quad y > 0, \tag{5.2.15}$$

then the condition (iii) concerning the continuity of φ across the gap produces

$$\psi(y) = 0 \quad \text{for } y \in \overline{L}. \tag{5.2.16}$$

Using the relation (5.2.15) in the representation (5.2.12), we find that $A(k)$ must satisfy

$$\frac{d^2}{dy^2} \int_0^\infty \frac{2A(k)}{k^2}\, M(k,y)\, dk = 2R\, e^{-Ky} + \psi(y), \quad y > 0$$

which is equivalent to

$$R\, e^{-Ky} + \int_0^\infty A(k)\, M(k,y)\, dk = -\frac{1}{2}\, \psi(y), \quad y > 0. \tag{5.2.17}$$

By Havelock's inversion we obtain that

$$A(k) = -\frac{1}{\pi(k^2 + K^2)} \int_L \psi(u)\, M(k,u)\, du, \tag{5.2.18}$$

$$R = -K \int_L \psi(u)\, e^{-Ku}\, du. \tag{5.2.19}$$

Again, using the condition (iii) in the representations (5.2.12), we obtain

$$\frac{d^2}{dy^2} \int_0^\infty \frac{A(k)}{k} M(k,y)\, dk = iK(1-R)e^{-Ky}, \quad y \in L. \quad (5.2.20)$$

By using the relation (5.2.18) in the left side of (5.2.20), we obtain, after interchanging the orders of integration,

$$\frac{d^2}{dy^2} \int_L \psi(u) \left\{ \int_0^\infty \frac{M(k,y)\,M(k,u)}{k(k^2+K^2)}\, dk \right\} du = -i\pi K(1-R)e^{-Ky}, \quad y \in L. \quad (5.2.21)$$

The inner integral can be evaluated (cf. Ursell (1947)), and thus we obtain

$$\frac{d^2}{dy^2} \int_L \mathcal{K}(u,y)\, \psi(u)\, du = -2i\pi K\, e^{-Ky}, \quad y \in L, \quad (5.2.22)$$

after using the relation (5.2.19) for R, where

$$\mathcal{K}(u,y) = \ln\left|\frac{y+u}{y-u}\right| - 2\, e^{-K(y+u)} \int_{-\infty}^{K(y+u)} \frac{e^v}{v}\, dv + 2i\pi\, e^{-K(y+u)}, \quad y,u \in L. \quad (5.2.23)$$

If we now interchange the order differentiation and integration in (5.2.22), we obtain formally

$$\int_L \mathcal{L}(u,y)\, \psi(u)\, du = -2i\pi K\, e^{-Ky}, \quad y \in L, \quad (5.2.24)$$

where

$$\mathcal{L}(u,y) = \frac{1}{(y-u)^2} + \frac{2K}{y+u} + \frac{1}{(y+u)^2} - 2K\, e^{-K(y+u)} \int_{-\infty}^{K(y+u)} \frac{e^v}{v}\, dv + 2i\pi K^2\, e^{-K(y+u)}. \quad (5.2.25)$$

Because of the presence of the hypersingular term $(y-u)^{-2}$ in the kernel $\mathcal{L}(u,y)$, the integral in the left side of (5.2.24) is not an ordinary one, and is to be interpreted in the sense of Hadamard finite part. Thus we arrive at the hypersingular integral equation of the form

$$\int_L \left[\frac{1}{(y-u)^2} + \mathcal{L}_1(y,u) \right] \psi(u)\, du = f(y), \quad y \in L$$

for the determination of $\psi(y)$, $\mathcal{L}_1(y,u)$ being regular and $f(y)$ being known.

Remark

As in the Problem 1, here also we again remark that, instead of the hypersingular integral equation (5.2.24), we can derive a *weakly singular* integral equation for $\psi(y)$, involving two arbitrary constants in the right side, as follows:

If we integrate both sides of the equation (5.2.22) twice, we obtain the integral equation

$$\int_L \mathcal{K}(u,y)\,g(u)\,du = -\frac{2i\pi}{K}\,e^{-Ky} + C_0 + C_1 y, \quad y \in L \quad (5.2.26)$$

in which C_0 and C_1 are two arbitrary constants, and $\mathcal{K}(y,u)$, given by the expression (5.2.23) is a kernel involving *logarithmic singularity* at $u = y$, so that the integral equation (5.2.26) is *weakly* singular. The two arbitrary constants C_0 and C_1 can be determined for particular configurations of the barrier. For example, if the barrier extends infinitely downwards from a fixed depth below the free surface, then $C_1 \equiv 0$ for the boundedness of the right side as $y \to \infty$.

Problem 4: A straight pressurized crack in an unbounded solid (the Griffith crack problem)

We consider the problem of a simple straight crack along the x-axis ($y = 0$, $a < x < b$) in a plane isotropic elastic medium under a pressure distribution $p(x)$ along both its edges. The problem is described as follows in the usual notations

$$\sigma_{xy} = 0 \text{ on } y = 0, \quad -\infty < x < \infty,$$

$$\sigma_{yy} = -p(x) \text{ on } y = 0, \quad a < x < b,$$

$$v = 0 \text{ on } y = 0, \quad -\infty < x < a, \quad b < x < \infty, \quad (5.2.27)$$

$$\sigma_{xx} \to 0, \quad \sigma_{xy} \to 0, \quad \sigma_{yy} \to 0 \text{ as } \left(x^2 + y^2\right)^{1/2} \to \infty.$$

For two-dimensional elasticity problems, the displacement components u, v and the stress components $\sigma_{xx}, \sigma_{xy}, \sigma_{yy}$ can be expressed by Muskhelishvili's formulae given by

$$2\mu(u + iv) = K\Phi(z) - z\,\overline{\Phi'(z)} - \overline{\Psi(z)},$$

$$\sigma_{xx} = \mathrm{Re}\left[\Phi'(z) + \overline{\Phi'(z)} - z\,\overline{\Phi''(z)} - \overline{\Psi'(z)}\right],$$

$$\sigma_{yy} = \mathrm{Re}\left[\Phi'(z) + \overline{\Phi'(z)} + z\overline{\Phi''(z)} + \overline{\Psi'(z)}\right], \quad (5.2.28)$$

$$\sigma_{xy} = \mathrm{Im}\left[\bar{z}\,\Phi''(z) + \Psi'(z)\right]$$

where

$$\mu = \frac{E}{2(1+\gamma)}, \quad K = 3 - 4\gamma, \quad (5.2.29)$$

E being the Young's modulus, γ being the Poisson's ratio, μ being the shear modulus, $z = x + iy$, $\Phi(z)$, $\Psi(z)$ being two arbitrary analytic functions.

Let

$$\rho(z) = z\,\Phi'(z) + \Psi(z), \quad (5.2.30)$$

then the formulae (5.2.28) take the forms

$$2\mu(u+iv) = K\Phi(z) - \overline{\rho(z)} + (\overline{z}-z)\overline{\Phi'(z)},$$

$$\sigma_{xx} = \mathrm{Re}\left[\Phi'(z) + 2\overline{\Phi'(z)} - \overline{\rho'(z)} + (\overline{z}-z)\overline{\Phi''(z)}\right],$$

$$\sigma_{yy} = \mathrm{Re}\left[\Phi'(z) + \overline{\rho'(z)} - (\overline{z}-z)\overline{\Phi''(z)}\right], \quad (5.2.31)$$

$$\sigma_{xy} = -\mathrm{Im}\left[\Phi'(z) - \rho'(z) - (\overline{z}-z)\overline{\Phi''(z)}\right]$$

where now $\Phi(z)$ and $\rho(z)$ are also two arbitrary analytic functions.

For the Griffith crack problem, we note that, if we assume $\Phi(z) = \rho(z)$, then it is ascertained from the very beginning that

$$\sigma_{xy} = 0 \quad \text{on} \quad y = 0.$$

In this case, the displacement components u, v and the stress components $\sigma_{xx}, \sigma_{yy}, \sigma_{xy}$ can be expressed as

$$2\mu(u+iv) = K\Phi(z) - \overline{\Phi(z)} + (\overline{z}-z) + \overline{\Phi'(z)}$$

$$\sigma_{xx} = \mathrm{Re}\left[\Phi'(z) + \overline{\Phi'(z)} + (\overline{z}-z)\overline{\Phi''(z)}\right],$$

$$\sigma_{yy} = \mathrm{Re}\left[\Phi'(z) + \overline{\Phi'(z)} - (\overline{z}-z)\overline{\Phi''(z)}\right], \quad (5.2.32)$$

$$\sigma_{xy} = -\mathrm{Im}\left[(z-\overline{z})\overline{\Phi''(z)}\right]$$

where now $\Phi(z)$ is an analytic function in the complex z-plane cut along the x-axis from a to b (the crack).

Let $\varphi(x)$ denote the displacement (unknown) of the points of the upper edge of the Griffith crack, then $-\varphi(x)$ is the corresponding displacement of the points of the lower edge of the crack. Thus

$$v(x,\pm 0) = \begin{cases} \pm\ \varphi(x), & a < x < b \\ 0 & , \quad \text{otherwise} \end{cases} \tag{5.2.33}$$

where $\varphi(x)$ is unknown, but $\varphi(a) = 0 = \varphi(b)$.

Now from the first equation of (5.2.32), we find

$$v(x,\pm 0) = \frac{1}{2\mu}\ \text{Im}\Big[K\Phi(z) - \overline{\Phi(z)} \Big]_{y=\pm 0}$$

$$= \frac{K+1}{2\mu}\ \text{Im}\ \Phi(z).$$

Let

$$\Phi(z) = \alpha(x,y) + i\beta(x,y).$$

Then

$$v(x,\pm 0) = \frac{K+1}{2\mu}\ \beta(x,\pm 0). \tag{5.2.34}$$

However (5.2.33) is satisfied if $v(x,y) = -v(x,-y)$ so that $\beta(x,y)$ can be chosen to satisfy $\beta(x.y) = -\beta(x,-y)$. Hence $\beta(x,y)$ is odd in y. But as $\dfrac{\partial\alpha}{\partial x} = \dfrac{\partial\beta}{\partial y}$, we see that $\alpha(x,y)$ is even in y. Thus $\alpha(x,y) = \alpha(x,-y)$ and $\beta(x,y) = -\beta(x,-y)$. Hence we must have that

$$\Phi(z) = \overline{\Phi(\bar z)}. \tag{5.2.35}$$

Since $\Phi(z)$ is sectionally analytic in the complex z-plane cut along the x-axis between a to b, let us represent it as

$$\Phi(z) = \int_a^b \frac{q(t)}{t-z}\ dt. \tag{5.2.36}$$

Then by Plemelj formulae,

$$\Phi(x\pm i0) = \pm\ \pi i q(x) + \int_a^b \frac{q(t)}{t-x}\ dt,\quad a < x < b, \tag{5.2.37}$$

so that

$$\Phi(x+i0) - \Phi(x-i0) = 2\pi i q(x), \quad a < x < b.$$

This gives

$$\alpha(x,+0) + i\beta(x,+0) - \alpha(x,-0) - i\beta(x,-0) = 2\pi i q(x), \quad a < x < b.$$

Since $\alpha(x,y)$ is even in y and $\beta(x,y)$ is odd in y, this is equivalent to

$$2i\beta(x,+0) = 2\pi i q(x), \quad a < x < b.$$

Using (5.2.34), we find from this

$$q(x) = \frac{1}{\pi} \beta(x,+0)$$

$$= \frac{2\mu}{\pi(K+1)} v(x,+0)$$

$$= \frac{2\mu}{\pi(K+1)} \varphi(x).$$

Thus $\Phi(z)$ is determined in terms of the unknown function $\varphi(x)$ in the form

$$\Phi(z) = \frac{2\mu}{\pi(K+1)} \int_a^b \frac{\varphi(t)}{t-z} dt. \qquad (5.2.38)$$

To obtain an integral equation for $\varphi(t)$, we use the condition

$$\sigma_{yy} = -p(x) \text{ on } y = 0, \quad a < x < b.$$

Now from the third equation of (5.2.33), we have on $y = 0$

$$\sigma_{yy} = \text{Re}\left[\Phi'(z) + \overline{\Phi'(z)}\right]_{y=0}$$

$$= \frac{4\mu}{\pi(K+1)} \int_a^b \frac{\varphi(t)}{(t-x)^2} dt.$$

Thus we see that $\varphi(t)$ satisfies the simple hypersingular integral equation

$$\int_a^b \frac{\varphi(t)}{(t-x)^2} dt = -\frac{\pi(K+1)}{4\mu} p(x), \quad a < x < b \qquad (5.2.39)$$

with the requirement that $\varphi(a) = \varphi(b) = 0$.

We shall obtain the exact solution of a simple hypersingular integral equation in the next section.

5.3 SOLUTION OF SIMPLE HYPERSINGULAR INTEGRAL EQUATION

We consider the simple hypersingular integral equation

$$(H\varphi)(x) = \frac{1}{\pi} \int_{-1}^{1} \frac{\varphi(t)}{(t-x)^2} \, dt = f(x), \quad -1 < x < 1 \quad (5.3.1)$$

with   the   end   conditions $\varphi(1) = \varphi(-1) = 0$,   where $f(x)$ is a known function and φ is required to be such that $\varphi \in C^{1,\alpha}(-1,1)$ where $C^{1,\alpha}(-1,1)$ denotes the class of functions having Hölder continuous derivatives with exponent $\alpha(0 < \alpha < 1)$ in the open interval $(-1,1)$ (cf. Martin (1992)).

(A) *Some elementary methods*

Let $(T\varphi)(x)$ denote the Cauchy singular operator

$$(T\varphi)(x) = \frac{1}{\pi} \int_{-1}^{1} \frac{\varphi(t)}{t-x} \, dt, \quad -1 < x < 1. \quad (5.3.2)$$

The hypersingular operator $(H\varphi)(x)$ can be interpreted as

$$H\varphi = \frac{d}{dx}(T\varphi). \quad (5.3.3)$$

Also, it is easy to show that

$$H\varphi = T\varphi' \quad (5.3.4)$$

where φ' denotes the derivative of φ. Thus the simple hypersingular integral equation (5.3.1) can be recast into the following two equivalent but basically different forms

$$\frac{d}{dx}(T\varphi) = f(x), \quad -1 < x < 1, \quad (5.3.5)$$

$$T\varphi' = f(x), \quad -1 < x < 1. \quad (5.3.6)$$

It then becomes clear that the solution of the HSIE (5.3.1) can be determined successfully, by solving any one of the Cauchy type singular integral equations (5.3.5) and (5.3.6). It may be noted that the problems of solving the SIEs (5.3.5) and (5.3.6) are basically different in the sense that the function φ in (5.3.5) has the end behaviour that $\varphi(\pm 1) = 0$ (in fact $\varphi(t) = 0\left(|t \mp 1|^{1/2}\right)$ as $t \to \pm 1$ while the function φ' in (5.3.6) has the behaviour that $\varphi'(t) = 0\left(|t \mp 1|^{-1/2}\right)$ as $t \to \pm 1$.

We now describe three basically independent methods of solution of the HSIE (5.3.1) (cf. Chakrabarti and Mandal (1998)).

Method 1

Let
$$f(x) = g'(x) \tag{5.3.7}$$

And using the equivalent representation (5.3.5) of the equation (5.3.1), we get

$$(T\varphi)(x) = B + g(x), \quad -1 < x < 1 \tag{5.3.8}$$

where B is an arbitrary constant. The equation (5.3.8) is a Cauchy type SIE satisfying the end conditions $\varphi(\pm 1) = 0$. Its solution is given by

$$\varphi(x) = -\frac{\left(1 - x^2\right)^{1/2}}{\pi} \int_{-1}^{1} \frac{B + g(t)}{\left(1 - t^2\right)^{1/2} (t - x)} \, dt, \quad -1 < x < 1, \tag{5.3.9}$$

provided

$$\int_{-1}^{1} \frac{B + g(t)}{\left(1 - t^2\right)^{1/2}} \, dt = 0 \tag{5.3.10}$$

which is the solvlability condition for SIE (5.3.8). Thus the arbitrary constant in (5.3.8) is given by

$$B = -\frac{1}{\pi} \int_{-1}^{1} \frac{g(t)}{\left(1 - t^2\right)^{1/2}} \, dt. \tag{5.3.11}$$

Since

$$\int_{-1}^{1} \frac{1}{\left(1 - t^2\right)^{1/2} (t - x)} \, dt = 0 \quad \text{for} \ -1 < x < 1$$

we find from (5.3.9) that

$$\varphi(x) = -\frac{(1-x^2)^{1/2}}{\pi} \int_{-1}^{1} \frac{g(t)}{(1-t^2)^{1/2}(t-x)} dt, \quad -1 < x < 1. \quad (5.3.12)$$

Although (5.3.12) gives the required solution, it is expressed in terms of g. However, we would like to obtain the solution in terms of f. For this, we use the following algebraic identity

$$\frac{1}{(1-t^2)^{1/2}(t-x)} = \frac{1}{(1-x^2)} \left[\frac{t+x}{(1-t^2)^{1/2}} + \frac{(1-t^2)^{1/2}}{t-x} \right]. \quad (5.3.13)$$

Now using the result for the indefinite integral

$$\int \frac{(1-t^2)^{1/2}}{t-x} dt = -(1-t^2) + x \sin^{-1} t - (1-x^2)^{1/2} K(x,t) + D(x) \quad (5.3.14)$$

where

$$K(x,t) = \ln \left| \frac{t-x}{1-tx + \{(1-t^2)(1-x^2)\}^{1/2}} \right|, \quad (5.3.14)$$

and $D(x)$ is an arbitrary function of x, we find that

$$\int \frac{1}{(1-t^2)^{1/2}(t-x)} dt = \frac{1}{(1-x^2)^{1/2}} \{K(x,t) + E(x)\} \quad (5.3.15)$$

where $E(x)$ is an arbitrary function of x.

Integrating the right side of the relation (5.3.12) by parts and using the result (5.3.15), we find that $\varphi(x)$ is given by

$$\varphi(x) = \frac{1}{\pi} \int_{-1}^{1} g'(t) K(x,t) dt = \frac{1}{\pi} \int_{-1}^{1} f(t) \ln \left| \frac{t-x}{1-tx + \{(1-t^2)(1-x^2)\}^{1/2}} \right| dt. \quad (5.3.16)$$

Method 2

In this method we utilize the equivalent representation (5.3.6) of the hypersingular integral equation (5.3.1). Then, the solution of the Cauchy type integral equation (5.3.6) with the requirement that $\varphi'(t) = 0\left(\left|1 \mp t\right|^{-1/2}\right)$ as $t \to \pm 1$, is given by

$$\varphi'(x) = \frac{1}{\left(1 - x^2\right)^{1/2}}\left[C - \frac{1}{\pi}\int_{-1}^{1}\frac{\left(1 - t^2\right)^{1/2} f(t)}{t - x}\,dt\right], \quad -1 < x < 1, \qquad (5.3.17)$$

where C is an arbitrary constant. Integrating both sides of (5.3.17) we obtain

$$\varphi(x) = D + C\,\sin^{-1}x + \frac{1}{\pi}\int_{-1}^{1}\left(1 - t^2\right)^{1/2} f(t)\left[\int \frac{dx}{\left(1 - x^2\right)^{1/2}(x - t)}\right]dt$$

where D is another constant. The inner integral in (cf.(5.3.15)

$$\frac{1}{\left(1 - t^2\right)^{1/2}}\left\{E(t) + K(t, x)\right\}$$

so that

$$\varphi(x) = D + C\,\sin^{-1}x + \frac{1}{\pi}\int_{-1}^{1} f(t)\,E(t)\,dt + \frac{1}{\pi}\int_{-1}^{1} f(t)\,K(t, x)\,dt$$

$$\qquad (5.3.18)$$

$$= D_0 + C\,\sin^{-1}x + \frac{1}{\pi}\int_{-1}^{1} f(t)\,K(x, t)\,dt$$

where D_0 is another constant.

Finally we observe that if $\varphi(x)$ has to satisfy the end conditions $\varphi(\pm 1) = 0$, we must choose $C = 0$, $D_o = 0$ and in that case the solution (5.3.18) reduces to (5.3.16), i.e.

$$\varphi(x) = \frac{1}{\pi}\int_{-1}^{1} f(t)\,\ln\left|\frac{t - x}{1 - tx + \left\{(1 - t^2)(1 - x^2)\right\}^{1/2}}\right|dt, -1 < x < 1.$$

Method 3

In the third method of solution of the hypersingular integral equation (5.3.1), along with the end conditions $\varphi(\pm 1) = 0$, we employ a direct approach, which leads to the problem of solving a simple Abel type integral equation.

By using the transformations

$$t = 2s - 1, \quad x = 2y - 1, \quad \varphi(t) = \varphi(2s - 1) = \psi(s), \quad f(x) = f(2y - 1) = g(y), \quad (5.3.19)$$

the hypersingular integral equation (5.3.1) reduces to

$$\frac{1}{\pi} \int_0^1 \frac{\psi(s)}{(s - y)^2} ds = 2g(y), \quad 0 < y < 1 \qquad (5.3.20)$$

with the end conditions $\psi(0) = \psi(1) = 0$.

We set

$$\psi(s) = s^{1/2} \int_s^1 \frac{S(\xi)}{(\xi - s)^{1/2}} d\xi, \quad 0 < s < 1 \qquad (5.3.21)$$

where $S(\xi)$ is a differentiable function in $(0,1)$, with $S(1) \neq 0$. We then find, by integrating by parts, that

$$\psi(s) = 2\{s(1-s)\}^{1/2} S(1) - 2s^{1/2} \int_s^1 (\xi - s)^{1/2} S'(\xi) \, d\xi, \quad 0 < s < 1. \quad (5.3.22)$$

Clearly, $\psi(s)$ belongs to the desired class of functions in which we seek the solution of the equation (5.3.20).

Substituting the relation (5.3.22) into the left side of the equation (5.3.20) and interchanging the order of integration in the repeated integral we obtain

$$\frac{2S(1)}{\pi} \int_0^1 \frac{\{s(1-s)\}^{1/2}}{(s-y)^2} ds - \frac{2}{\pi} \int_0^1 S'(\xi) \left\{ \int_0^\xi \frac{\{s(\xi - s)\}^{1/2}}{(s - y)^2} ds \right\} d\xi = 2g(y), \quad 0 < y < 1. \quad (5.3.23)$$

We now use the following results, which can be easily derived, by using elementary integrations

$$\int_0^1 \frac{\{s(\xi - s)\}^{1/2}}{s - y} ds = \begin{cases} -\pi \left(y - \dfrac{\xi}{2} \right) & \text{for } 0 < y < \xi, \\[2mm] \pi \left[\dfrac{\xi}{2} - y + \{y(y - \xi)\}^{1/2} \right] & \text{for } y > \xi, \end{cases} \qquad (5.3.24)$$

where for $0 < y < \xi$, the integral is in the sense of CPV. From (5.3.24), it follows after putting $\xi = 1$ that

$$\int_0^1 \frac{\{s(1-s)\}^{1/2}}{s - y} ds = -\pi \left(y - \frac{1}{2} \right) \quad \text{for } 0 < y < 1.$$

Hence the hypersingular integrals in (5.5.23) are evaluated. Thus the equation (5.3.23) finally produces

$$-2S(1)+2\int_0^1 S'(\xi)\,d\xi - \int_0^y S'(\xi)\,\frac{2y-\xi}{\{y(y-\xi)\}^{1/2}}\,d\xi = 2g(y),\quad 0<y<1. \quad (5.3.25)$$

Using the symbol

$$A(y)=\int_0^y (y-\xi)^{1/2}\,S'(\xi)\,d\xi \quad\quad (5.3.26)$$

the equation (5.3.25) can be expressed as

$$y^{1/2}A'(y)+\frac{1}{2y^{1/2}}A(y)=-S(0)-g(y),\quad 0<y<1 \quad (5.2.27)$$

where $A'(y)=\dfrac{dA}{dy}$.

The equation (5.3.27) is a first-order differential equation for the function $A(y)$ and solving it with the initial condition $A(0)=0$, as is obvious from the definition (5.3.26), we easily determine $A(y)$, from which we derive that

$$2A'(y)\equiv\int_0^y \frac{S'(\xi)}{(y-\xi)^{1/2}}\,d\xi = G(y),\quad 0<y<1 \quad (5.3.28)$$

where

$$G(y)=y^{-3/2}[-S(0)y+\int_0^y g(\eta)\,d\eta - 2yg(y)] \quad (5.3.29)$$

We have thus arrived at the simple Abel type integral equation (5.3.28) for the determination of the unknown function $S'(\xi)$, and from the solution (2.1.9b) in Chapter 2, we find that

$$S'(\xi)=\frac{1}{\pi}\,\frac{d}{d\xi}\int_0^\xi \frac{G(y)}{(\xi-y)^{1/2}}\,dy,\quad 0<\xi<1,$$

giving on integration,

$$S(\xi)=S(0)+\frac{1}{\pi}\int_0^\xi \frac{G(y)}{(\xi-y)^{1/2}}\,dy,\quad 0<\xi<1. \quad (5.3.30)$$

Substituting for $G(y)$ giving in (5.3.29) and interchanging the orders of the repeated integrals, we obtain

$$S(\xi) = S(0) - \frac{S(0)}{\pi} \int_0^\xi \frac{dy}{\{y(\xi - y)\}^{1/2}} - \frac{2}{\pi} \int_0^\xi \frac{g(y)}{\{y(\xi - y)\}^{1/2}} dy$$

$$+ \frac{1}{\pi} \int_0^\xi g(y) \left\{ \int_y^\xi \frac{d\eta}{\eta^{3/2}(\xi - s)^{1/2}} \right\} dy, \quad 0 < \xi < 1. \tag{5.3.31}$$

Using the results, obtained by elementary means,

$$\int_0^\xi \frac{dy}{\{y(\xi - y)\}^{1/2}} = \pi$$

and

$$\int_y^\xi \frac{d\eta}{\eta^{3/2}(\xi - \eta)^{1/2}} = \frac{2}{\xi} \left(\frac{\xi - y}{y} \right)^{1/2},$$

we obtain, from the relation (5.3.31), that

$$S(\xi) = -\frac{2}{\pi \xi} \int_0^\xi \frac{y^{1/2} g(y)}{(\xi - y)^{1/2}} dy, \quad 0 < \xi < 1. \tag{5.3.32}$$

Substituting the final form (5.3.32) of the function $S(\xi)$ into the relation (5.3.21), we find that

$$\psi(s) = -\frac{2}{\pi} s^{1/2} \int_s^1 \frac{1}{\xi(\xi - s)^{1/2}} \left\{ \int_0^\xi \frac{y^{1/2} g(y)}{(\xi - y)^{1/2}} dy \right\} d\xi$$

$$= -\frac{2}{\pi} s^{1/2} \left[\int_0^s y^{1/2} g(y) \left\{ \int_s^1 \frac{d\xi}{\xi \{(\xi - s)(\xi - y)\}^{1/2}} \right\} dy \right. \tag{5.3.33}$$

$$\left. + \int_s^1 y^{1/2} g(y) \left\{ \int_y^1 \frac{d\xi}{\xi \{(\xi - s)(\xi - y)\}^{1/2}} \right\} dy \right]$$

after interchanging the order of integration. The inner integrals in the two terms of the right side of the relation (5.3.33) can be evaluated by using elementary methods, and we find that

$$\int_{\max(y,s)}^1 \frac{d\xi}{\xi \{(\xi - s)(\xi - y)\}^{1/2}} = \frac{1}{(sy)^{1/2}} \ln \left| \frac{s + y - 2sy + 2\{sy(1-s)(1-y)\}^{1/2}}{s - y} \right|. \tag{5.3.34}$$

Using (5.3.34) in (5.3.33) we find finally

$$\psi(s) = \frac{2}{\pi} \int_0^1 g(y) \ln \left| \frac{s-y}{s+y-2sy+2\{sy(1-s)(1-y)\}^{1/2}} \right|, \quad 0<s<1 \quad (5.3.35)$$

Substituting back in terms of the original variables from (5.3.19), i.e.

$$s = \frac{t+1}{2}, \ y = \frac{x+1}{2}, \ \psi\left(\frac{t+1}{2}\right) = \varphi(t), \ g\left(\frac{x+1}{2}\right) = f(x), \quad \text{we obtain}$$

$$\varphi(t) = \frac{1}{\pi} \int_{-1}^1 f(x) \ln \left| \frac{x-t}{1-xt+\{(1-x^2)(1-t^2)\}^{1/2}} \right| dt, \ -1<t<1. \quad (5.3.36)$$

(B) Function-theoretic method

It is observed that even though the original integral equation (5.3.1) involves hypersingular kernel, its solution as given by the formula (5.3.36) involves a singular integral possessing Cauchy type singularity. This is in fact *weaker* compared to the *strong* singularity of the given integral equation. Because of the fact that once hypersingular integrals have been accepted, there is no special need to deviate and bring in integrals with weaken singularities in the picture.

Chakrabarti (2007) developed a direct *function-theoretic* method to solve the hypersingular integral equation (5.3.1) by reducing it to a Riemann-Hilbert type boundary value problem of an unknown sectionally analytic function of a complex variable $z(=x+iy)$, in the complex z-plane, cut along the segment $(-1,1)$ of the real axis. This method is now presented here.

Let us define $\Phi(z)$ as

$$\Phi(z) = \int_{-1}^1 \frac{\varphi(t)}{(t-z)^2} \, dt, \ z \notin (-1,1). \quad (5.3.37)$$

Then $\Phi(z)$ is a sectionally analytic function of z in the complex z-plane cut along $(-1,1)$ of the real axis.

We have the result (cf. Jones (1982), p 104)

$$\lim_{y \to \pm 0} \frac{1}{x+iy} = \mp i\pi\delta(x) + \frac{1}{x}, \quad -\infty<x<\infty \quad (5.3.38)$$

where $\delta(x)$ denotes Dirac delta function. Differentiation of both sides with respect to x produces formally

$$\lim_{y \to \pm 0} \frac{1}{(x+iy)^2} = \pm i\pi\delta'(x) + \frac{1}{x^2} \qquad (5.3.39)$$

where $\delta'(x)$ denote the derivative of $\delta(x)$.

Utilizing (5.3.39) in (5.3.37) we obtain the following Plemelj-type formulae for the limiting values of the function, as z approaches a point on the cut $(-1,1)$ from above $(y \to +0)$ and below $(y \to -0)$ respectively:

$$\Phi(x \pm i0) \equiv \Phi^\pm(x)$$

$$= \mp i\pi \int_{-1}^{1} \varphi(t)\, \delta'(t-x)\, dt + \int_{-1}^{1} \frac{\varphi(t)}{(t-x)^2}\, dt, \quad -1 < x < 1.$$

so that

$$\Phi^\pm(x) = \pm i\pi\varphi'(x) + \int_{-1}^{1} \frac{\varphi(t)}{(t-x)^2}\, dt, \quad -1 < x < 1. \qquad (5.3.40)$$

It may be noted that the limiting values (5.3.40) can also be derived by *standard* Plemelj formulae involving the limiting values of the Cauchy type integral

$$\Psi(z) = \int_{-1}^{1} \frac{\varphi(t)}{t-z}\, dt, \quad z \notin (-1,1) \qquad (5.3.41)$$

giving

$$\Psi(x \pm i0) \equiv \Psi^\pm(x)$$

$$= \pm i\pi\, \varphi(x) + \int_{-1}^{1} \frac{\varphi(t)}{t-x}\, dt, \quad -1 < x < 1 \qquad (5.3.42)$$

where the integral is in the sense of CPV, and noting

$$\Phi(z) = \frac{d\Psi}{dz}(z) \qquad (5.3.43)$$

along with

$$\int_{-1}^{1} \frac{\varphi(t)}{(t-x)^2}\, dt = \frac{d}{dx} \int_{-1}^{1} \frac{\varphi(t)}{t-x}\, dt, \quad -1 < x < 1. \qquad (5.3.44)$$

Now, the two relations in (5.3.40) can be viewed as the following two equivalent relations

$$\Phi^+(x) + \Phi^-(x) = 2 \int_{-1}^{1} \frac{\varphi(t)}{(t-x)^2} \, dt, \quad -1 < x < 1, \quad (5.3.45a)$$

$$\Phi^+(x) - \Phi^-(x) = 2\pi i \, \varphi'(x), \quad -1 < x < 1. \quad (5.3.45b)$$

Using the relation (5.3.45a), the hypersingular integral equation (5.3.1) reduces to

$$\Phi^+(x) + \Phi^-(x) = 2\pi f(x), \quad -1 < x < 1 \quad (5.3.46)$$

which represents a *special Riemann-Hilbert type* boundary value problem for the determination of the unknown function $\Phi(z)$. Its solution can be found directly as explained below.

If $\Phi_0(z)$ represents a nontrivial solution of the *homogeneous* problem (5.3.46), satisfying

$$\Phi_0^+(x) + \Phi_0^-(x) = 0, \quad -1 < x < 1 \quad (5.3.47)$$

then we may rewrite the nonhomogeneous problem (5.3.46) as

$$\Omega^+(x) - \Omega^-(x) = \frac{2\pi f(x)}{\Phi_0^+(x)}, \quad -1 < x < 1 \quad (5.3.48)$$

where

$$\Omega(z) = \frac{\Phi(z)}{\Phi_0(z)}. \quad (5.3.49)$$

Thus, the relation (5.3.45b) suggests that we can determine the function $\Omega(z)$ as

$$\Omega(z) = \frac{1}{2\pi i} \int_{-1}^{1} \frac{g(t)}{(t-z)^2} \, dt + E(z) \quad (5.3.50)$$

where

$$g'(x) = \frac{2\pi f(x)}{\Phi_0^+(x)}, \quad (5.3.51)$$

and $E(z)$ is an entire function of z.

Using the form (5.3.50) of the function $\Omega(z)$, along with the relation (5.3.49) and the Plemelj-type formula (5.3.45b). we obtain

$$\varphi'(x) = -\frac{1}{2\pi^2} \Phi_0^+(x) \left[\int_{-1}^{1} \frac{g(t)}{(t-x)^2} dt + 2\pi i \, E(x) \right], \quad -1 < x < 1. \tag{5.3.52}$$

Thus we have been able to determine the first derivative of the unknown function $\varphi(x)$, in terms of an unknown $E(x)$, $E(z)$ being an entire function in the complex z-plane.

If we select $\Phi_0(z)$ as

$$\Phi_0(z) = \left(\frac{z+1}{z-1} \right)^{1/2}, \tag{5.3.53}$$

then

$$\Phi_0^\pm(x) = \mp \, i \left(\frac{1+x}{1-x} \right)^{1/2} \quad \text{for } -1 < x < 1. \tag{5.3.54}$$

Now, by definition, $|\Phi(z)| = 0\left(\dfrac{1}{|z|^2} \right)$ as $|z| \to \infty$, so that $|\Omega(z)| = \left| \dfrac{\Phi(z)}{\Phi_0(z)} \right| = 0\left(\dfrac{1}{|z|^2} \right)$ as $|z| \to \infty$, and hence we must select $E(z) \equiv 0$.

Thus from (5.3.51) and (5.3.54) we find

$$\varphi'(x) = -\frac{1}{\pi} \left(\frac{1+x}{1-x} \right)^{1/2} \int_{-1}^{1} \frac{h(t)}{(t-x)^2} dt, \quad -1 < x < 1 \tag{5.3.55}$$

where

$$h'(t) = \left(\frac{1-t}{1+t} \right)^{1/2} f(t). \tag{5.3.56}$$

Finally, integrating the relation (5.3.55), the solution of the given integral equation (5.3.1) is obtained as

$$\varphi(x) = p(x) + C, \quad -1 < x < 1 \tag{5.3.57}$$

where

$$p'(x) = -\frac{1}{\pi}\left(\frac{1+x}{1-x}\right)^{1/2}\int_{-1}^{1}\frac{h(t)}{(t-x)^2}\,dt \qquad (5.3.58)$$

and C is an arbitrary constant. We note that the expression for $h(t)$ will involve an arbitrary constant of integration, arising out of the relation (5.3.56), and thus the form of the solution (5.3.57) involves *two* arbitrary constants altogether.

This completes the function-theoretic method of solution of the hypersingular integral equation (5.3.1) in which the conditions $\varphi(\pm1) = 0$ have not been used. When these conditions are used, the form of the solution (5.3.57) will reduce to the form (5.3.36). This is now shown below.

Integration of (5.3.58) by parts produces

$$p'(x) = -\frac{1}{\pi}\left(\frac{1+x}{1-x}\right)^{1/2}\int_{-1}^{1}\frac{h'(t)}{t-x}\,dt + \frac{1}{\pi}\left[\frac{h(-1)-h(1)}{\left(1-x^2\right)^{1/2}} + \frac{2h(1)}{\left(1-x\right)\left(1-x^2\right)^{1/2}}\right].$$

Another integration gives, because of the relation (5.3.56),

$$p(x) = \frac{1}{\pi}\int_{-1}^{1} f(t)\,ln\left|\frac{x-t}{1-xt+\left\{(1-x^2)(1-t^2)\right\}^{1/2}}\right|\,dt \qquad (5.3.59)$$

$$+\ 4\pi\ h(1)\ \frac{1+(1-x^2)^{1/2}}{1-x+(1-x^2)^{1/2}} + \frac{1}{\pi}\left[h(-1)-h(1)+\int_{-1}^{1}\left(\frac{1-t}{1+t}\right)^{1/2}f(t)\,dt\right]\sin^{-1}x$$

ignoring an arbitrary constant and using the following results

$$\left(\frac{1-t}{1+t}\right)^{1/2}\frac{1}{t-x} = \frac{\left(1-t^2\right)^{1/2}}{1+x}\left(\frac{1}{t-x}-\frac{1}{1+t}\right) \qquad (5.3.60)$$

and

$$\frac{\partial}{\partial x}\,ln\left|\frac{x-t}{1-xt+\left\{(1-x^2)(1-t^2)\right\}^{1/2}}\right| = \left(\frac{1-t^2}{1-x^2}\right)^{1/2}\frac{1}{x-t}. \qquad (5.3.61)$$

When $\varphi(\pm1) = 0$, we must have $h(1) = 0 = C$, and

$$h(-1) = -\int_{-1}^{1}\left(\frac{1-t}{1+t}\right)^{1/2}f(t)\,dt \qquad (5.3.62)$$

and thus we obtain

$$\varphi(x) = \frac{1}{\pi} \int_{-1}^{1} f(t) \ln \left| \frac{x-t}{1-xt + \left\{ (1-x^2)(1-t^2) \right\}^{1/2}} \right| dt, \quad -1 < x < 1,$$

which agrees with (5.3.36).

5.4 SOLUTION OF HYPERSINGULAR INTEGRAL EQUATION OF THE SECOND KIND

In this section we present a straightforward analysis involving the complex function theoretic method to determine the closed form solution of a special hypersingular integral equation of the second kind. This analysis is given by Chakrabarti, Mandal, Basu and Banerjea (1997). The hypersingular integral equation of the second kind is given by

$$\varphi(x) - \frac{\alpha}{\pi} \left(1 - x^2 \right)^{1/2} \int_{-1}^{1} \frac{\varphi(t)}{(t-x)^2} dt = f(x), \quad -1 < x < 1 \quad (5.4.1)$$

with $\varphi(\pm 1) = 0$, and it is known as the elliptic wing case of Prandtl's equation (cf. Dragos (1994a, 1994b), for which $\alpha(>0)$ is a known constant, and $f(x) = \frac{2\pi k}{\beta} \left(1 - x^2 \right)^{\frac{1}{2}}$; k, β also being known constants with $\beta > 0$. The equation (5.4.1) is examined here for its closed form solution for the class of forcing functions $f(x)$ which are summable in $(-1,1)$, by using a complex function theoretic method in a straight forward manner.

The present method converts the equation (5.4.1) into a *differential* Riemann-Hilbert Problem on the cut $(-1,1)$ which is amenable to its closed form solution, and, by this, it is demonstrated that in the considered special case the elaborate reduction method as presented in the treatise of Muskhelishvilli (1953, formulae (122.13)-(122.15) and (122.17)) can be avoided to solve the given integral equation.

Assuming that $\varphi \in C^{1,\gamma}(-1,1)$, i.e. $\varphi(x)$ possesses a Hölder continuous first derivative, with exponent γ, such that $0 < \gamma < 1$, so that $\varphi(x)$ is summable in $(-1,1)$, and $\varphi(\pm 1) = 0$, we set

$$\Phi(z) = \frac{1}{2\pi i} \int_{-1}^{1} \frac{\varphi(t)}{t-z} dt, \quad z \notin (-1,1), \quad (5.4.2)$$

and observe that $\Phi(z)$ represents a sectionally analytic function of the complex variable $z(=x+iy)$ in the z-plane, cut along the segment $(-1,1)$ on the real axis, and that $|\varphi(z)| = 0\left(\dfrac{1}{|z|}\right)$ as $|z| \to \infty$.

The Plemelj formulae give that

$$\Phi^{\pm}(x) = \pm\frac{1}{2}\,\varphi(x) + \frac{1}{2\pi i}\int_{-1}^{1}\frac{\varphi(t)}{t-x}\,dt, \quad -1 < x < 1 \qquad (5.4.3)$$

where $\Phi^{\pm}(x)$ are the limiting values of $\Phi(z)$ on the upper and lower sides of the cut $(-1,1)$ and the integral is in the sense of CPV.

Using the formulae (5.4.3), we can easily cast the integral equation (5.4.1) into the functional relation as given by

$$\left(1 - i\alpha(1-x^2)^{1/2}\,\frac{d}{dx}\right)\Phi^{+}(x) - \left(1 + i\alpha(1-x^2)^{1/2}\,\frac{d}{dx}\right)\Phi^{-}(x) = f(x), \quad -1 < x < 1. \ (5.4.4)$$

The relation (5.4.4) represents a *differential* Riemann-Hilbert problem on the cut (cf. Gakhov (1966)) for the determination of the function $\Phi(z)$, and once the function $\Phi(z)$ is determined, the solution $\varphi(x)$ of the equation (5.4.1) can be completed by using the formulae (5.4.3) once again, giving

$$\varphi(x) = \Phi^{+}(x) - \Phi^{-}(x), \quad -1 < x < 1. \qquad (5.4.5)$$

Now we determine the function $\Phi(z)$, satisfying the relation (5.4.4), by using the analysis as described below.

We observe that the following limiting values hold for the sectionally analytic function $(z^2 - 1)^{1/2}$ in the complex z-plane, cut along $(-1,1)$ on the real axis:

$$\lim_{y \to \pm 0}(z^2 - 1)^{1/2} = \pm i(1-x^2) \qquad (5.4.6)$$

with that branch of the square root for which $x^{1/2}$ is positive whenever x is positive. Then the relation (5.4.4) can be cast into the simplified form given by

$$\Psi^{+}(x) - \Psi^{-}(x) = f(x), \quad -1 < x < 1, \qquad (5.4.7)$$

where

$$\Psi(z) = \Phi(z) - \alpha\left(z^2 - 1\right)^{1/2}\frac{d\Phi(z)}{dz}. \qquad (5.4.8)$$

The solution of the simple Riemann-Hilbert problem (5.4.7) is obtained, immediately, in the form

$$\psi(z) = \frac{1}{2\pi i} \int\limits_{-1}^{1} \frac{f(t)}{t-z} \, dt, \qquad (5.4.9)$$

since $|\psi(z)| = 0(\frac{1}{|z|})$ as $|z| \to \infty$ because $|\Phi(z)| = 0(\frac{1}{|z|})$ as $|z| \to \infty$ (cf. the representation (5.4.6)).

Then, solving the ordinary differential equation (5.4.8) by a standard method, we obtain that (since $\alpha > 0$),

$$\Phi(z) = \left[-\frac{1}{\alpha} \int\limits_{\infty}^{z} \frac{\left\{\varsigma + (\varsigma^2-1)^{1/2}\right\}^{-\frac{1}{\alpha}}}{(\varsigma^2-1)^{1/2}} \Psi(\varsigma) \, d\varsigma + \lambda \right] \left\{z + (z^2-1)^{1/2}\right\}^{\frac{1}{\alpha}} (5.4.10)$$

where λ is an arbitrary constant.

We note that we must choose $\lambda = 0$, in order to meet with the requirement that $|\Phi(z)| = 0(\frac{1}{|z|})$ as $|z| \to \infty$. We also note that, since analytic functions of analytic functions are themselves analytic in the same cut plane, there is no difficulty in checking that the function as given by (5.4.10) represents an analytic function in the complex z-plane, cut along the segment $(-1,1)$ of the real axis. In particular, by Cahchy's integral theorem applied in the outer region of the cut $(-1,1)$, $\Phi(z)$ is a single-valued function there.

We thus finally determine the function $\Phi(z)$, related to the integral equation (5.4.1), in the form

$$\Phi(z) = -\frac{1}{\alpha} \left\{z + (z^2-1)^{1/2}\right\}^{1/\alpha} \int\limits_{\infty}^{z} \frac{\left\{\varsigma + (\varsigma^2-1)^{1/2}\right\}^{-1/\alpha}}{(\varsigma^2-1)^{1/2}} \Psi(\varsigma) \, d\varsigma \quad (5.4.11)$$

where $\Psi(z)$ is as given by the formulae (5.4.9). As $z \to \pm 1$, we have $\Phi^+(\pm 1) = \Phi^-(\pm 1)$ so that $\varphi(\pm 1) = 0$, as required.

The result (5.4.11), along with the relation (5.4.5), thus completely solves the integral equation (5.4.1).

In the case of Prandlt's equation, we have

$$\alpha = \frac{2\pi}{\beta} \ (\beta > 0), \ \text{and} \ f(x) = \frac{2\pi k}{\beta} (1-x^2)^{1/2}. \quad (5.4.12)$$

The integral in the relation (5.4.9) can be evaluated by standard contour integral equation procedure (cf. Gakhov (1966) for example), and we find that

$$\Psi(z) = -\frac{\pi k}{i\beta} \left\{ z + (z^2 - 1)^{1/2} \right\}^{-1}.$$ (5.4.13)

Using the result (5.4.13), along with the value of α as $\alpha = \dfrac{2\pi}{\beta}$, we determine the function $\Phi(z)$, by using the formula (5.4.11), after some easy manipulations, in the form

$$\Phi(z) = \frac{2ik}{1 + \dfrac{2}{\pi}\beta} \left\{ z - (z^2 - 1)^{1/2} \right\}.$$ (5.4.14)

The relation (5.4.5) finally determines the solution $\varphi(x)$ of the integral equation of Prandtl, and obtains that

$$\varphi(x) = \frac{4\pi k}{1 + \dfrac{2}{\pi}\beta} \left(1 - x^2\right)^{1/2},$$ (5.4.15)

and this completely agrees with the one as quoted in the Dragos's papers.

The second kind hypersingular integral equation (5.4.1) will be considered later for its numerical solution.

Chapter 6

Singular Integro-differential Equations

In this chapter we consider some singular integro-differential equations for their solution by employing a simplified analysis. These equations arise in a natural way while solving a class of mixed boundary value problems of mathematical physics (cf. Holford (1964), Spence (1960), Stewartson (1960)).

6.1 A CLASS OF SINGULAR INTEGRO-DIFFERENTIAL EQUATIONS

We consider the singular integro-differential equation as given by

$$u(x) - \frac{1}{\pi} \int_0^\infty \frac{v(t)}{t-x} \, dt = f(x), \quad 0 < x < \infty \tag{6.1.1}$$

where $u(x)$ and $v(x)$ represent two linear differential expressions of the forms

$$u(x) = \sum_{k=0}^n a_k \, \varphi^{(k)}(x) \tag{6.1.2}$$

and

$$v(x) = \sum_{k=0}^n b_k \, \varphi^{(k)}(x) \tag{6.1.3}$$

in which a_k's and b_k's are in general complex constants, $\varphi^{(k)}(x)$ denotes the kth derivative of the unknown function $\varphi(x)$ with prescribed initial values $\phi^{(k)}(0)$, along with the conditions that $\phi^{(k)}(\infty) = 0$, for $k = 0, 1, 2, ..., n$, and $f(x)$ represents a known differentiable function.

Varley and Walker (1989) have discussed a general method of solving the integro-differential equation (6.1.1) by converting it to a singular integral equation, which avoids the complication of calculating various singular integrals appearing in the final form of the solution. Chakrabarti and Sahoo (1996) employed a straightforward analysis simplifying the work of Verley and Walker (1989). This is explained here.

We seek solution of (6.1.1) under the assumptions that

$$v(x) = 0(x^{\alpha}) \text{ as } x \to 0 \tag{6..1.4}$$

where $\text{Re } \alpha > -1$, and

$$v(x) = 0(x^{\beta}) \text{ as } x \to \infty \tag{6..1.5}$$

where $-1 < \text{Re } \beta < 0$.

Method of solution

Using the Laplace transform $V(p)$ of the function $v(x)$ defined by

$$V(p) = \int_0^\infty v(x) e^{-px} dx, \quad p > 0, \tag{6.1.6}$$

the equation (6.1.1) can be shown to be equivalent to the following singular integral equation

$$U(p) + \frac{1}{\pi} \int_0^\infty \frac{V(q)}{q - p} dq = F(p) \tag{6.1.7}$$

where $U(p)$, $V(p)$ and $F(p)$ are the Laplace transforms of the functions $u(x)$, $v(x)$ and $f(x)$ respectively. The expressions for $U(p)$ and $V(p)$ are given by

$$U(p) = A(p)\Phi(p) - A_1(p), \quad V(p) = B(p)\Phi(p) - B_1(p) \tag{6.1.8}$$

where

$$A(p) = \sum_{k=0}^{n} a_k \, p^k$$

$$A_1(p) = \left(\sum_{k=1}^{n} a_k \, p^{k-1} \right) \varphi_0 + \left(\sum_{k=2}^{n} a_k \, p^{k-2} \right) \varphi_1 + \cdots + a_n \varphi_{n-1},$$

$$B(p) = \sum_{k=0}^{n} b_k \, p^k, \qquad\qquad (6.1.9)$$

$$B_1(p) = \left(\sum_{k=1}^{n} b_k \, p^{k-1} \right) \varphi_0 + \left(\sum_{k=2}^{n} b_k \, p^{k-2} \right) \varphi_1 + \cdots b_n \varphi_{n-1}$$

with

$$\varphi_k = \varphi^{(k)}(0), \quad k = 0, 1, 2, \dots n.$$

Using the relations (6.1.8), the equation (6.1.7) can be cast into the form

$$A(p)V(p) + \frac{B(p)}{\pi} \int_0^\infty \frac{V(q)}{q-p} \, dq = C(p), \quad p > 0 \qquad (6.1.10)$$

where

$$C(p) = A_1(p)B(p) - B_1(p)A(p) + F(p)B(p). \qquad (6.1.11)$$

By the Abelian theorem on Laplace transforms (cf. Doetsch (1974)) and using the assumptions (6.1.4) and (6.1.5), we obtain

$$V(p) = 0(p^{-1-\alpha}) \text{ as } p \to \infty. \qquad (6.1.12)$$

and

$$V(p) = 0(p^{-1-\beta}) \text{ as } p \to 0. \qquad (6.1.13)$$

These orders of $V(p)$ assure that the integral in (6.1.10) exists and finite.

We now discuss, in some detail, the method of determination of the function $V(p)$. We consider the general case of (6.1.1) for which the pair

(a_0, b_0) has the property that either a_0 and b_0 or both are non-zero, the same is assumed to hold good for the pair (a_n, b_n) also.

Defining

$$\Psi(z) = \frac{1}{2\pi i} \int_0^\infty \frac{V(t)}{t - z} \, dt, \quad z = p + ip', \tag{6.1.14}$$

we note that $\Psi(z)$ is analytic in the complex z-plane with a cut along the positive real axis. Then by Plemelj formulae

$$\Psi^+(p) - \Psi^-(p) = V(p), \quad p > 0 \tag{6.1.15}$$

and

$$\Psi^+(p) + \Psi^-(p) = \frac{1}{i\pi} \int_0^\infty \frac{V(t)}{t - p} \, dt, \quad p > 0 \tag{6.1.16}$$

where $\Psi^+(p)$ and $\Psi^-(p)$ are the limiting values of $\Psi(z)$ as z approaches a point p on the positive real axis from above and from below respectively.

From (6.1.10) and the relations (6.1.15) and (6.1.16) we obtain

$$\frac{A(p) + iB(p)}{A(p) - iB(p)} \Psi^+(p) - \Psi^-(p) = \frac{C(p)}{A(p) - iB(p)}, \quad p > 0, \tag{6.1.17}$$

which represents a Riemann-Hilbert problem for the determination of the sectionally analytic function $\Psi(z)$. This Riemann-Hilbert problem can be solved as described below.

We first construct a sectionally analytic function $\Omega(z)$ in the complex z-plane cut along the positive real axis, satisfying the relation

$$\frac{\Omega^+(p)}{\Omega^-(p)} = \frac{A(p) + iB(p)}{A(p) - iB(p)}, \quad p > 0 \tag{6.1.18}$$

where $\Omega^+(p)$, $\Omega^-(p)$ are the limiting values of $\Omega(z)$ on the two sides of the positive real axis.

Assuming that $A(p) + iB(p)$ possesses zeros at the points μ_i $(i = 1, 2, ...n)$ and $A(p) - iB(p)$ possesses zeros at *different* points λ_j $(j = 1, 2, n)$, none of which lie on the positive real axis, and setting

$$\Omega(z) = \Lambda(z)\, z^{\delta} \tag{6.1.19}$$

with

$$\delta = \frac{1}{2\pi i}\; ln\; \frac{a_n + i b_n}{a_n - i b_n}, \tag{6.1.20}$$

we can recast the problem (6.1.18) in the form

$$\frac{\Lambda^+(p)}{\Lambda^-(p)} = \prod_{j=1}^{n} \frac{p - \mu_j}{p - \lambda_j}, \quad p > 0. \tag{6.1.21}$$

The solution of the problem posed by the relation (6.1.21) can easily be written down as (cf. Gakhov (1966)),

$$ln\; \Lambda(z) = \frac{1}{2\pi i} \int_0^{\infty} \left(\sum_{j=1}^{n} ln\; \frac{t - \mu_j}{t - \lambda_j} \right) \frac{1}{t - z}\, dt\; + ln\; E_1(z) \tag{6.1.22}$$

where $E_1(z)$ is an entire function, and using the Plemelj formulae, we obtain

$$ln\, \frac{E_1(p)}{\Lambda^+(p)} = \sum_{j=1}^{n} \left[\frac{1}{2}\, ln\, \frac{p - \lambda_j}{p - \mu_j} + \frac{1}{2\pi i} \int_0^{\infty} ln\left(\frac{t - \lambda_j}{t - \mu_j} \right) \frac{1}{t - p}\, dt \right], \quad p > 0. \tag{6.1.23}$$

After simplification this gives

$$\frac{E_1(p)}{\Lambda^+(p)} = \prod_{j=1}^{n} p^{\gamma_j} (p - \lambda_j) V_j(p) \tag{6.1.24}$$

where

$$V_j(p) = c_j \left(\frac{p - \lambda_j}{p - \mu_j} \right)^{\frac{\theta_{2j}}{2\pi}} (p - \lambda_j)^{-(1+\gamma_j)} \exp\left[-\frac{\lambda_j - \mu_j}{2\pi i} \int_0^{p} \frac{ln\left(\frac{t}{|\mu_j|} \right)}{(t - \lambda_j)(t - \mu_j)}\, dt \right] \tag{6.1.25}$$

with

$$\gamma_j = -\frac{1}{2\pi i}\; ln\left(\frac{\lambda_j}{\mu_j} \right), \quad \theta_{1j} = \arg(\lambda_j),\quad \theta_{2j} = \arg(\mu_j),$$

$$c_j = \exp\left[-\frac{\lambda_j - \mu_j}{2\pi i} I(0) + \gamma_j \{ln\, |\lambda_j| - i(2\pi - \theta_{1j} - \theta_{2j})\}\right]$$

and

$$I(0) = \int_0^\infty \frac{ln\, t}{(t-\lambda_j)(t-\mu_j)} dt.$$

We thus determine the limiting value of $\dfrac{1}{\Omega^+(p)}$, as given by

$$\frac{1}{\Omega^+(p)} = \frac{p^{\delta+\sum_{j=1}^{n}\gamma_j} \prod_{j=1}^{n} \{(p-\lambda_j)V_j(p)\}}{E_1(p)}. \tag{6.1.26}$$

Using the relation (6.1.18), we can recast (6.1.17) as

$$\Omega^+(p)\Psi^+(p) - \Omega^-(p)\Psi^-(p) = \frac{C(p)\Omega^-(p)}{A(p)-iB(p)}, \quad p > 0, \tag{6.1.27}$$

producing

$$\Omega(z)\Psi(z) = \frac{1}{2\pi i} \int_0^\infty \frac{C(t)\Omega^-(t)}{A(t)-iB(t)} \frac{1}{t-z} dt + E_2(z), \tag{6.1.28}$$

where $E_2(z)$ is an entire function.

By Plemelj formulae, we obtain from (6.1.28)

$$\Psi^+(p) = \frac{1}{2} \frac{C(p)}{A(p)-iB(p)} \frac{\Omega^-(p)}{\Omega^+(p)} + \frac{1}{2\pi i\Omega^+(p)} \int_0^\infty \frac{C(t)\Omega^-(t)}{A(t)-iB(t)} \frac{1}{t-p} dt$$
$$+ \frac{E_2(p)}{\Omega^+(p)}, p > 0 \tag{6.1.29}$$

and

$$\Psi^-(p) = -\frac{1}{2} \frac{C(p)}{A(p)-iB(p)} + \frac{1}{2\pi i\Omega^-(p)} \int_0^\infty \frac{C(t)\Omega^-(t)}{A(t)-iB(t)} \frac{1}{t-p} dt + \frac{E_2(p)}{\Omega^-(p)}, p > 0. \tag{6.1.30}$$

The function $V(p)$ is finally determined, by using the relations (6.1.24) and (6.1.30) in the relation (6.1.15), and we find that

$$V(p) = \frac{1}{2} \frac{C(p)}{A(p)-iB(p)} \left(\frac{\Omega^-(p)}{\Omega^+(p)} + 1 \right) + \frac{1}{2\pi i} \left(\frac{1}{\Omega^+(p)} - \frac{1}{\Omega^-(p)} \right) \int_0^\infty \frac{C(t)\Omega^-(t)}{A(t)-iB(t)} \frac{1}{t-p} dt$$

$$+ \left(\frac{1}{\Omega^+(p)} - \frac{1}{\Omega^-(p)} \right) E_2(p). \tag{6.1.31}$$

The integral in the relation (6.1.31) can be easily evaluated by considering the contour integral

$$\int_\Gamma \frac{C(\zeta)\Omega(\zeta)}{B(\zeta)(\zeta - z)} d\zeta$$

where Γ is closed contour comprising of a circle of large radius along with a loop around the positive real axis in the complex ζ-plane, with the assumption that $\zeta_1, \zeta_2, \dots \zeta_n$ are the n distinct zeros of $B(\zeta)$, which do not lie on the positive real axis and $B'(\zeta_j) \neq 0$ for $j = 1, 2, \dots n$ (the case of multiple zeros can also be dealt with). Then, by application of the Cauchy residue theorem we obtain that (using the relation (6.1.18) also)

$$\frac{2}{\pi} \int_0^\infty \frac{C(t)\Omega^-(t)}{A(t)-iB(t)} \frac{1}{t-p} dt = \frac{C(p)}{B(p)} \Omega^+(p) \left(1 + \frac{\Omega^-(p)}{\Omega^+(p)} \right) + 2 \sum_{j=1}^n \frac{\Omega(\zeta_j)C(\zeta_j)}{B'(\zeta_j)(\zeta_j - p)}. \tag{6.1.32}$$

Using the result (6.1.31) in (6.1.32) we derive that

$$V(p) = \frac{B(p) \left[\sum_{j=1}^n \frac{\Omega(\zeta_j)C(\zeta_j)}{B'(\zeta_j)(p-\zeta_j)} - 2iE_2(p) \right]}{(a_n - ib_n)(p-\lambda_1)(p-\lambda_2)\cdots(p-\lambda_n)\Omega^+(p)}, \tag{6.1.33}$$

which, on substitution for $\Omega^+(p)$ from the relation (6.1.26), gives

$$V(p) = H(p) \, p^{\delta+\gamma} \, V_1(p)V_2(p)\cdots V_n(p) \tag{6.1.34}$$

where

$$H(p) = \frac{B(p) \left[\sum_{j=1}^n \frac{\Omega(\zeta_j)C(\zeta_j)}{B'(\zeta_j)(p-\zeta_j)} - 2i\, E_2(p) \right]}{(a_n - ib_n)E_1(p)} \tag{6.1.35}$$

which is a rational function, with

$$\gamma = \gamma_1 + \gamma_2 + \cdots + \gamma_n.$$

The analysis that has been explained above is applicable to (6.1.1) for the general case where a_k's and b_k's are complex constants. But, for the particular case, when a_k's and b_k's are real constants, we have $0 < \delta < 1$. Setting $\theta_{1j} = \theta_j$, $\theta_{2j} = 2\pi - \theta_j$, and $\mu_j = \bar{\lambda}_j$, we obtain

$$\gamma_j = 1 - \frac{\theta_j}{\pi}, \quad c_j = 1,$$

and using the relation (6.1.25) we find that

$$V_j(p) = \frac{\exp\left[-\frac{\sin\theta_j}{\pi} \int_0^{\frac{p}{|\lambda_j|}} \frac{\ln t}{t^2 - 2t \cos\theta_j + 1} dt \right]}{\left[(p - \lambda_j)(p - \bar{\lambda}_j) \right]^{1 - \frac{\theta_j}{2\pi}}}. \tag{6.1.36}$$

This is equivalent to the result obtained by Varley and Walker (1989).

Ultimately, by using the form (6.1.34) we find that

$$V(p) = 0\left(\frac{H(p)}{p^{n-\delta}} \right) \quad \text{as } p \to \infty \tag{6.1.37}$$

and

$$V(p) = 0\left(H(p)p^{\gamma+\delta} \right) \quad \text{as } p \to 0, \tag{6.1.38}$$

which help is expressing $H(p)$ as

$$H(p) = \sum_{j=-k}^{n-1} h_j p^j \tag{6.1.39}$$

where $k - 1 < \delta + \gamma < k$, $k = 0, 1, ..., (n-1)$, when the relations (6.1.12) and (6,1,13) are also used.

The determination of h_j's can be completed by considering two separate cases as described next. These cases take care of even different degrees of the two polynomials $A(p)$ and $B(p)$.

Case (i): If $A(p)$ and $B(p)$ are of the same degree n, the constants $h_0, h_1, ..., h_{n-1}$ can be obtained by using the relation (6.1.10) along with the fact that $\zeta_1, \zeta_2, ... \zeta_n$ are the simple zeros of $B(p)$. The other constants $h_{-1}, h_{-2}, ... h_{-n+1}$ will remain arbitrary.

Case (ii): If $n > m$ where n is the degree of $A(p)$ and m is the degree of $B(p)$, then the m constants $h_1, h_2, ..., h_m$ can be determined by using the m zeros of $B(p)$ and the relation (6.1.10). The other $(n - m - 1)$ constants $h_{m+1}, ..., h_{n-1}$ can be determined by using the asymptotic behaviour of $\Omega(z)$ as $z \to \infty$. The other constants will remain arbitrary.

The function $v(x)$ can finally be determined by using the Laplace inversion formula and we find that

$$v(x) = 2 \sum_{j=0}^{n} R_j \cos\left(x|\lambda_j|\sin\theta_j + \alpha_j\right) e^{x|\lambda_j|\cos\theta_j} + \int_0^\infty l(s) \, e^{-xs} \, ds \quad (6.1.40)$$

where $R_j e^{i\alpha_j}$ denotes the residue of $V(p)$ at $p = \lambda_j$ and

$$l(s) = \frac{1}{2\pi i} \left\{ V(se^{-i\pi}) - V(se^{i\pi}) \right\} \quad (6.1.41)$$

with the expression for $V(p)$ as given by the relation (6.1.34). The unknown function $\varphi(x)$ can then be determined successfully, by using the second relation in (6.1.8) along with the convolution theorem for Laplace transforms.

Remark

Followings are some observations on the method that has been presented above.

(i) A lot of complication is avoided by evaluating the Cauchy type integral in the relation (6.1.31) directly.

(ii) In this analysis, the determination of $H(p)$ is simpler as compared to that explained by Varley and Walker (1989).

(iii) The behaviour of the funtion $V(p)$ at the end points $p = 0$ and $p = \infty$ along with the analyticity property of $V(p)$ has helped to arrive at the expression for $H(p)$, as given by the relation (6.1.35).

6.2 A SPECIAL TYPE SINGULAR INTEGRO-DIFFERENTIAL EQUATION

In this section we describe a method of solution of a special type of singular integro-differential equation arising in the study of a problem concerning heat conduction and radiation, in elastic contact problems, etc. (cf. Frankel (1995), Chakrabarti and Hamsapriye (1999)). This equation is given by

$$\mu \frac{d\theta(x)}{dx} - \lambda \int_0^1 \frac{\theta(y)}{y-x} \, dy = f(x), \quad 0 < x < 1 \qquad (6.2.1)$$

where λ, μ are known non-zero constants, and $f(x)$ is a known differentiable function of $x \in (0,1)$, with the end onditions for the unknown function θ being given by

$$\theta(0) = \theta(1) = 0.$$

Using the transformations

$$\lambda = \frac{1}{2}(\xi + 1), \quad y = \frac{1}{2}(\eta + 1), \quad \Theta(\xi) = \theta\left(\frac{\xi+1}{2}\right) \qquad (6.2.2)$$

we can rewrite the equation (6.1.1) as

$$2\mu \frac{d\Theta}{d\xi} - f\left(\frac{\xi+1}{2}\right) = \lambda \int_{-1}^1 \frac{\Theta(\eta)}{\eta - \xi} \, d\eta, \quad -1 < \xi < 1 \qquad (6.2.3)$$

with the end conditions

$$\Theta(\pm 1) = 0.$$

If we use the inversion formula for the Cauchy type singular integral equation, as given by the relation (6.2.3), assuming that the left side of the relation is known, for the time being, and that we require bounded solutions at the end points $\xi = \pm 1$ in the form $\Theta(\pm 1) = 0$, we find that

$$\Theta(\xi) = -\frac{2\mu}{\lambda\pi^2}(1-\xi^2)^{1/2} \int_{-1}^1 \frac{1}{(1-\eta^2)^{1/2}} \left\{ \frac{d\Theta}{d\eta} - \frac{1}{2\mu} f\left(\frac{\eta+1}{2}\right) \right\} \frac{1}{\eta-\xi} \, d\eta, -1 < \xi < 1, \quad (6.2.4)$$

with the condition for bounded solution being given by

$$\int_{-1}^1 \frac{1}{(1-\eta^2)^{1/2}} \left\{ \frac{d\Theta}{d\eta} - \frac{1}{2\mu} f\left(\frac{\eta+1}{2}\right) \right\} d\eta = 0. \qquad (6.2.5)$$

If in the relations (6.2.4) and (6.2.5) we use the transformation

$$\Theta(\xi) = (1 - \xi^2) \, \varphi(\xi) \qquad (6.2.6)$$

where φ is non-zero at $\xi = \pm 1$, the transformation being motivated by the equation (6.2.3) under the end conditions for $\Theta(\xi)$, we obtain the following relations:

$$\left(1-\xi^2\right)^{1/2}\varphi(\xi) = -\frac{2\mu}{\pi^2}\int_{-1}^{1}\left\{\left(1-\eta^2\right)^{1/2}\varphi'(\eta) - \frac{2\eta}{\left(1-\eta^2\right)^{1/2}}\varphi(\eta)\right.$$

$$\left. - \frac{1}{2\mu(1-\eta^2)^{1/2}}f\left(\frac{\eta+1}{2}\right)\right\}\frac{1}{\eta-\xi}\,d\eta, -1<\xi<1 \tag{6.2.7}$$

and

$$\int_{-1}^{1}\left\{\left(1-\eta^2\right)^{1/2}\varphi'(\eta) - \frac{2\eta}{\left(1-\eta^2\right)^{1/2}}\varphi(\eta) - \frac{1}{2\mu(1-\eta^2)^{1/2}}f\left(\frac{\eta+1}{2}\right)\right\}d\eta = 0. \tag{6.2.8}$$

Now, in order to solve the equation (6.2.7), we assume that $\varphi(\xi)$ can be expressed in terms of a convergent series as given by

$$\varphi(\xi) = \sum_{n=0}^{\infty} a_n\, T_n(\xi) \tag{6.2.9}$$

where $T_n(\xi)$ $(n = 0,1,2,...)$ are Chebyshev polynomials of the first kind, and a_n's are unknown constants to be determined. Then using the following results on Chebyshev polynomials $T_n(\xi)$ and $U_n(\xi)$ of first and second kinds

(i) $T_n'(\xi) = n\, U_{n-1}(\xi),\ \ n = 1,2,...,$

(ii) $T_0'(\xi) = 0,$

(iii) $T_m(\xi)T_n(\xi) = \frac{1}{2}\left\{T_{m+n}(\xi) + T_{|m-n|}(\xi)\right\}, \ \ m,n = 0,1,2,...,$

(iv) $T_1(\xi) = \xi,$

(v) $\displaystyle\int_{-1}^{1}\frac{\left(1-\xi^2\right)^{1/2}U_n(\xi)}{\xi-\eta}\,d\xi = -\pi\, T_{n+1}(\eta),\ \ \eta=0,1,2,...,\ |\eta|<1,$

(vi) $\displaystyle\int_{-1}^{1}\frac{T_{n+1}(\xi)}{\left(1-\xi^2\right)^{1/2}(\xi-\eta)}\,d\xi = \pi\, U_n(\eta),\ \ n=0,1,2,...,\ |\eta|<1,$

(vii) $\displaystyle\int_{-1}^{1} \frac{\xi\, T_n(\xi)}{\left(1-\xi^2\right)^{1/2}}\,d\xi = \begin{cases} 0 \text{ if } n \neq 1, \\[2mm] \dfrac{\pi}{2} \text{ if } n = 1, \end{cases}$

the equation (6.2.7) produces, after substituting the relation (6.2.9) into it,

$$\left(1-\xi^2\right)^{1/2} \sum_{n=0}^{\infty} a_n\, T_n(\xi) = P(\xi) + \frac{2\mu}{\pi} \sum_{n=0}^{\infty} n\, a_n\, T_n(\xi) \qquad (6.2.10)$$

$$+ \frac{2\mu}{\lambda\pi} \sum_{n=0}^{\infty} a_n \left\{ U_n(\xi) + U_{|n-2|}(\xi) \right\}$$

where

$$P(\xi) = \frac{1}{\lambda\pi^2} \int_{-1}^{1} \frac{f\left(\dfrac{\eta+1}{2}\right)}{\left(1-\eta^2\right)^{1/2}(\eta-\xi)}\,d\eta. \qquad (6.2.11)$$

Also the relation (6.2.8) gives

$$\sum_{n=0}^{\infty} n\, a_n \int_{-1}^{1} (1-\eta^2)^{1/2} U_{n-1}(\eta)\,d\eta - \pi a_1 = \frac{1}{2\mu} \int_{-1}^{1} \frac{f\left(\dfrac{n+1}{2}\right)}{\left(1-\eta^2\right)^{1/2}}\,d\eta. \qquad (6.2.12)$$

Now using the result

$$\int_{-1}^{1}\left(1-\eta^2\right)^{1/2} U_{n-1}(\eta)\,d\eta = \begin{cases} 0 \text{ if } n \neq 1, \\[2mm] \dfrac{\pi}{2} \text{ if } n = 1, \end{cases} \qquad (6.2.13)$$

the relation (6.2.12) determines the unknown constant a_1 as

$$a_1 = -\frac{1}{\pi\mu} \int_{-1}^{1} \frac{f\left(\dfrac{\eta+1}{2}\right)}{\left(1-\eta^2\right)^{1/2}}\,d\eta. \qquad (6.2.14)$$

Next, in order to determine the other constants $a_n\,(n \neq 1)$, we multiply both sides of the relation (6.2.10) by $\left(1-\xi^2\right)^{-1/2} T_m(\xi)$ and integrate between -1 to 1 to obtain

$$\sum_{n=0}^{\infty} a_n \int_{-1}^{1} T_n(\xi)T_m(\xi)d\xi = \int_{-1}^{1} \frac{P(\xi)T_m(\xi)}{(1-\xi^2)^{1/2}} d\xi + \frac{2\mu}{\lambda\pi} \sum_{n=0}^{\infty} n \, a_n \int_{-1}^{1} \frac{T_n(\xi)T_m(\xi)}{(1-\xi^2)^{1/2}} d\xi$$

$$+ \frac{2\mu}{\lambda\pi} \sum_{n=0}^{\infty} a_n \int_{-1}^{1} \frac{T_m(\xi)}{(1-\xi^2)^{1/2}} \left\{ U_n(\xi) + U_{|n-2|}(\xi) \right\} d\xi, \ m = 0,1,2,\dots . \quad (6.2.15)$$

In the relation (6.2.15) if we use the results

$$T_m(\xi)T_n(\xi) = \frac{1}{2} \left\{ T_{m+n}(\xi) + T_{|m-n|}(\xi) \right\}$$

and

$$\int_{-1}^{1} \frac{T_m(\xi)T_n(\xi)}{(1-\xi^2)^{1/2}} d\xi = \begin{cases} 0 & \text{if } m \neq n, \\ \pi & \text{if } m = n = 0, \\ \dfrac{\pi}{2} & \text{if } m = n > 0, \end{cases}$$

we obtain the following relations connecting the unknown constants a_n:

$$\sum_{n=0}^{\infty} a_n \, c_{n0} = \int_{-1}^{1} \frac{P(\xi)}{(1-\xi^2)^{1/2}} d\xi + \frac{2\mu}{\lambda\pi} \sum_{n=0}^{\infty} a_n \, b_{n0}, \ m = 1,2,\dots, \quad (6.2.16)$$

$$\sum_{n=0}^{\infty} a_n \, c_{nm} = \int_{-1}^{1} \frac{P(\xi)T_m(\xi)}{(1-\xi^2)^{1/2}} d\xi + \frac{\mu}{\lambda} m \, a_m + \frac{2\mu}{\lambda\pi} \sum_{n=0}^{\infty} a_n \, b_{nm}, \ m = 1,2,\dots, \quad (6.2.17)$$

where

$$c_{nm} = \int_{-1}^{1} T_n(\xi)T_m(\xi) \, d\xi$$

and

$$b_{nm} = \int_{-1}^{1} \frac{T_m(\xi)}{(1-\xi^2)^{1/2}} \left\{ U_n(\xi) + U_{|n-2|}(\xi) \right\} d\xi. \quad (6.2.18)$$

The unknown constants $a_n \, (n = 0,1,2,\dots)$ can now be determined, approximately by truncating the series for $n = N$ in the relations (6.2.16) and (6.2.17) and thereby obtaining a system of linear equations for $(N+1)$ unknown $a_0, a_1, \dots, a_N$, of which the constant a_1 has to be given by the relation (6.2.14) for consistency.

Having described a general method to solve approximately the singular integro-differential equation (6.2.3), we present in the next section *three* methods of solution given by Chakrabarti and Hamsapriya (1999). Convergence aspect of one method is also discussed.

6.3 NUMERICAL SOLUTION OF A SPECIAL SINGULAR INTEGRO-DIFFERENTIAL EQUATION

Here we present *three* numerical methods of solution of the special singular integro-differential equation

$$2\frac{d\varphi}{dx} - \lambda \int_{-1}^{1} \frac{\varphi(t)}{t-x} \, dt = f(x), \quad -1 < x < 1 \tag{6.3.1}$$

where λ is a known positive constant, and $\varphi(x)$ satisfies the end conditions $\varphi(\pm 1) = 0$, and $f(x)$ has the special form $f(x) = -x/2$ This equation arises in a heat conduction and radiation problem as already mentioned in section 6.2

Method 1

In the first method, a colloation method is employed instead of the Galerkin method of Frankel (1995), after recasting the equation (6.3.1) into the form

$$\varphi(x) = \frac{1}{\pi \left(1-x^2\right)^{1/2}} \left[c' - \frac{1}{\lambda\pi} \int_{-1}^{1} \frac{\left(1-t^2\right)^{1/2}}{t-x} \left(f(t) - 2\frac{d\varphi}{dt} \right) dt \right], \quad -1 < x < 1, \tag{6.3.2}$$

where c' is defined as

$$c' = \int_{-1}^{1} \varphi(t) \, dt.$$

In the special case when $f(x) = -\frac{x}{2}$, as considered by Frankel (1995),

the relation (6.3.2) can be transformed to a more convenient form as

$$\left(1-x^2\right)^{1/2}\varphi(x) - \frac{1}{\pi} \int_{-1}^{1} \varphi(t) \, dt - \frac{2}{\lambda\pi^2} \int_{-1}^{1} \frac{\left(1-t^2\right)^{1/2}}{t-x} \frac{d\varphi}{dt} \, dt = -\frac{T_2(x)}{4\lambda\pi}, -1 < x < 1 \tag{6.3.3}$$

after utilizing the result

$$\int_{-1}^{1} \frac{t\left((1-t^2)^{1/2}\right)}{t-x} \, dt = -\frac{\pi}{2} \, T_2(x), \quad -1 < x < 1. \qquad (6.3.4)$$

It must be emphasized that the right side of the equation (6.3.3) will be more complicated in the case of any general $f(x)$ (other than a polynomial itself). However, approximating $f(x)$ by a suitable polynomial, the methods described here are still applicable.

The relation (6.3.3) can be rewritten in the operator form as

$$\left(D + \frac{2}{\lambda\pi} \, C\right) \varphi(x) = F(x), \quad -1 < x < 1 \qquad (6.3.5)$$

where

$$(D\varphi)(x) = \left(1-x^2\right)^{1/2} \varphi(x) - \frac{1}{\pi} \int_{-1}^{1} \varphi(t) \, dt, \qquad (6.3.6)$$

$$(C\varphi)(x) = -\frac{1}{\pi} \int_{-1}^{1} \frac{\left(1-t^2\right)^{1/2}}{t-x} \frac{d\varphi}{dt}, \quad -1 < x < 1, \quad (6.3.7)$$

$$F(x) = -\frac{1}{4\lambda\pi} \, T_2(x). \qquad (6.3.8)$$

Now we assume an approximation for the function $\varphi(x)$ in the form

$$\varphi(x) \approx \varphi_N(x) \equiv \sum_{j=0}^{N} {}'' \, c_j^{(N)} \, T_j(x) \qquad (6.3.9a)$$

giving

$$\varphi'(x) = \sum_{j=0}^{N} {}'' \, c_j^{(N)} \, T_j'(x) = \sum_{j=0}^{N} {}'' \, c_j^{(N)} jU_{j-1}(x) \qquad (6.3.9b)$$

with the constants $c_j^{(N)}$ $(j = 0, 1, ..., N)$ are to be determined by the aid of a suitably selected set of collocation points, to be described below, for a sufficiently large value of the positive integer N, ensuring the convergence of the method used. Here double primes in the summation symbol denote that the first and the last terms are halved.

Substituting the approximations (6.3.9) into the relation (6.3.3) and using the results

$$\int_{-1}^{1} \frac{\left(1-t^2\right)^{1/2}}{t-x} \, U_{j-1}(x) = \pi \, T_j(x), \quad -1 < x < 1,$$

$$\int_{-1}^{1} T_j(t)dt = \frac{1+(-1)^j}{1-j^2}, \quad j=0,1,2,..., \tag{6.3.10}$$

we arrive at the relation

$$\sum_{j=0}^{N} {}'' c_j^{(N)} \left[\left(1-x^2\right)^{1/2} T_j(x) - \frac{1}{\pi} \frac{1+(-1)^j}{1-j^2} + \frac{2j}{\lambda\pi} T_j(x) \right] = -\frac{T_2(x)}{4\lambda\pi}, \quad -1 < x < 1. \tag{6.3.11}$$

We now select the set of collocation points as given by

$$x = x_i = -\cos\left(\frac{i\pi}{N}\right), \quad i = 1,2,...,(N-1),$$

and derive the system of linear equations

$$\sum_{j=0}^{N} {}'' c_j^{(N)} \left[\left(1-x_i^2\right)^{1/2} T_j(x_i) - \frac{1}{\pi} \frac{1+(-1)^j}{1-j^2} + \frac{2j}{\lambda\pi} T_j(x_i) \right] = -\frac{T_2(x_i)}{4\lambda\pi}, \quad i=1,2,...,(N-1). \tag{6.3.12}$$

Also, on using the end conditions $\varphi(\pm 1) = 0$, we obtain two more equations

$$\sum_{j=0}^{N} {}'' c_j^{(N)} = 0, \quad \sum_{j=0}^{N} {}'' (-1)^j \, c_j^{(N)} = 0. \tag{6.3.13}$$

We have therefore reduced the problem of solving the equation (6.3.3), to that of solving the system of $(N+1)$ linear equations given by (6.3.12) and (6.3.13) for the unknown constants $c_j^{(N)}$ ($j=0,1,2,...,N$), which can be handled by any standard method. The above system of linear equations is solved for $c_j^{(N)}$ ($j=0,1,2,...N$) and $\varphi_N(x)$ given by (6.3.9a) is computed at some intermediate points. These computed values of $\varphi_N(x)$ at some points are given in Table 1 (taken from Chakrabarti and Hamsapriye(1999)). We note that, because of the symmetry property $\varphi(x) = \varphi(-x)$, the $c_j^{(N)}$'s in the expression of $\varphi(x)$ have the property that $c_{2k+1}^{(N)} = 0$, $k = 0,1,...,[N/2]$.

We have considered three different values of $\lambda(>0)$ e.g. $\lambda = 0.1, 1, 10$ as considered by Frankel (1995), and have selected $N = 10, 20, 40$ and 80 for determining the constants $c_j^{(N)}$ we have also fixed the inermediate points to be $x = x_k = \pm 0.2k$, $k = 0, 1, ..., 5$, in all the methods. We have tabulated the values of $\varphi_N(x)$ only in the interval $[0,1]$, and symmetry takes care of the negative half of the interval $[-1,1]$.

The convergence of the numerical method is apparent from the given table. The numerical results so obtained tally with those of Frankel (1995).

Table 1

$\varphi_N(x)$ (Method 1)

λ	x_i / N	10	20	40	80
0.1	0	0.11578506	0.11578505	0.11578504	0.11578504
	0.2	0.11124202	0.11124204	0.11124204	0.11124204
	0.4	0.09757759	0.09757695	0.09757693	0.09757693
	0.6	0.07467364	0.07467364	0.07467375	0.07467375
	0.8	0.04230536	0.04230520	0.04230520	0.04230520
	1.0	0	0	0	0
1	0	0.06950789	0.06950780	0.06950773	0.06950773
	0.2	0.06712763	0.06712801	0.06712802	0.06712802
	0.4	0.05985113	0.05984643	0.05984624	0.05984626
	0.6	0.04718193	0.04718294	0.04718280	0.04718280
	0.8	0.02809934	0.02809826	0.02809828	0.02809828
	1.0	0	0	0	0
10	0	0.01392298	0.01392237	0.01392221	0.01392221
	0.2	0.01358400	0.01358756	0.01358769	0.01358769
	0.4	0.01255415	0.01253918	0.01253874	0.01253848
	0.6	0.01059876	0.01060232	0.01060181	0.01060179
	0.8	0.00727896	0.00727837	0.00727857	0.00727859
	1.0	0	0	0	0

Method 2

In the second method, we utilize the end-bounded solution of the integro-differential equation (6.3.1) in the form

$$\varphi(x) = -\frac{\left(1-x^2\right)^{1/2}}{\lambda \pi^2} \int_{-1}^{1} \left(f(t) - 2\frac{d\varphi}{dt} \right) \frac{1}{\left(1-t^2\right)^{1/2} (t-x)} \, dt, \quad -1 < x < 1, \quad (6.3.14)$$

which satisfies the end requirements $\varphi(\pm 1) = 0$, provided the condition

$$\int_{-1}^{1} \left(f(t) - 2\frac{d\varphi}{dt} \right) \frac{1}{\left(1-t^2\right)^{1/2}} dt = 0 \qquad (6.3.15)$$

is satisfied.

In fact, the formula (6.3.14) for $\varphi(x)$, can be shown to be equivalent to the formula (6,3,2), under the condition (6,3,15), beause of the fact that

$$\left(1-t^2\right)^{1/2} = \frac{1-t^2}{\left(1-t^2\right)^{1/2}} = \frac{\left(1-x^2\right)+\left(x^2-t^2\right)}{\left(1-t^2\right)^{1/2}}, \quad x,t \in (-1,1),$$

giving

$$\int_{-1}^{1} \frac{\left(1-t^2\right)^{1/2}\psi(t)}{x-t} dt = \left(1-x^2\right) \int_{-1}^{1} \frac{\psi(t)}{\left(1-t^2\right)^{1/2}(x-t)} dt + \int_{-1}^{1} \frac{t\,\psi(t)}{\left(1-t^2\right)^{1/2}} dt$$

$$+ x \int_{-1}^{1} \frac{\psi(t)}{\left(1-t^2\right)^{1/2}} dt \quad \text{for } \psi \in L(\rho), \quad -1 < x < 1,$$

where $L(\rho)$ denotes the space of all real square injtegrable functions on $(-1,1)$ with respect to the weight function $\rho(x) = (1-x^2)^{-1/2}$.

Then, the relation (6.3.14) follows by choosing $\psi(x) = f(x) - 2\dfrac{d\varphi}{dx}$ and c' to be

$$c' = -\frac{1}{\pi} \int_{-1}^{1} \frac{t\,\psi(t)}{\left(1-t^2\right)^{1/2}} dt \equiv \int_{-1}^{1} \varphi(t)\, dt.$$

It is rather obvious that the symmetry of $\varphi(x)$ as well as the antisymmetry of $f(x)$ automatically forces the condition (6.3.15) to be satisfied.

The relation (6.3.14) can be viewed as an equivalent integro-differential equation to the original equation (6.3.1) that satisfies the end conditions. An approximate solution to the new equation (6.3.14) has now been derived here by writing down the unknown function $\varphi(x)$ in terms of the Chebyshev approximation as given by

$$\varphi(x) \approx \varphi_N(x) = \sum_{j=0}^{N}{''} a_j^{(N)} T_j(x), \quad -1 < x < 1 \qquad (6.3.16)$$

where $a_i^{(N)}$ (j=0,1,...,N) are certain constants to be determined, which are different from the ones used in Method 1.

Substituting the approximation (6.3.16) into the equation (6.3.14) and using the result, for $-1 < x < 1$,

$$u_j(x) = \int_{-1}^{1} \frac{U_j(t)}{\left(1-t^2\right)^{1/2}(t-x)} dt = \begin{cases} 0 \text{ for } j = 0, -1, \\ 2\pi \text{ for } j = 1, \\ 2\pi \left\{U_{(j-1)/2}(x)\right\}^2 \text{ for } j = \text{odd integer}, \\ 2\pi \, U_{(j-2)/2}(x) \, U_{j/2}(x) \text{ for } j = \text{even integer}, \end{cases} \qquad (6.3.17)$$

we arrive at the relation

$$\sum_{j=0}^{N} {}'' a_j^{(N)} \left[T_j(x) - j \frac{2\left(1-x^2\right)^{1/2}}{\lambda \pi^2} u_{j-1}(x) \right] = \frac{\left(1-x^2\right)^{1/2}}{2\lambda \pi}, \quad -1 < x < 1, \qquad (6.3.18)$$

in the case when $f(x) = -\dfrac{x}{2}$.

We next discretize the relation in (6.3.18) at the Chebyshev points
 $x = x_i = -\cos\dfrac{i\pi}{N}$, $i = 0,1,...,N$ and obtain a system of $(N+1)$
linear equations in the $(N+1)$ unknowns $a_j^{(N)}$ $(j = 0,1,...,N)$, whose solution is then obtained by any standard method.

Numerical values of $\varphi_N(x)$, obtained by Method 2, are tabulated in Table 2 (taken from Chakrabarti and Hamsapriye (1999)), at the same set of intermediate points as are chosen for Method 1. It is clear that we have obtained the results as close as the results of Table 1. As a check to the solution (6.3.16) (i.e. $\varphi_N(x)$) thus obtained, we have verified that the relation (6.3.15) is satisfied automatically since $a_{2k+1}^{(N)} = 0$ for $k = 0,1,...,[N/2]$.

Method 3

We have also developed a Galerkin type method as discussed by Frankel (1995) using the reduced equation (6.3.14). The derivation is similar to that in Frankel (1995). We assume an expansion of $\varphi(x)$

$$\varphi(x) = \sum_{j=1}^{\infty} b_j \, T_{2j-2}(x) - 1, \quad -1 \le x \le 1. \qquad (6.3.19)$$

Using the prescribed end conditions $\varphi(\pm 1) = 0$, we obtain the relation

$$\sum_{j=1}^{\infty} b_j = 1 \qquad (6.3.20)$$

Using (6.3.20), we can rewrite the expression for $\varphi(x)$ in the form

$$\varphi(x) = \sum_{j=1}^{\infty} b_j \left(T_{2j-2}(x) - T_0(x) \right), \quad -1 \le x \le 1. \tag{6.3.21}$$

Substituting the relation (6.3.21) into the relation (6.2.14) and multiplying by the factor

Table 2

$\varphi_N(x)$ (Method 2)

λ	x_i / N	10	20	40	80
0.1	0	0.11578771	0.11578525	0.11578505	0.11578504
	0.2	0.11124464	0.11124224	0.11124205	0.11124204
	0.4	0.09758022	0.09757715	0.09757694	0.09757693
	0.6	0.07467610	0.07467397	0.07467371	0.07467375
	0.8	0.04230763	0.04230538	0.04230521	0.04230520
	1.0	0	0	0	0
1	0	0.06952008	0.06950874	0.06950781	0.06950774
	0.2	0.06713985	0.06712895	0.06712809	0.06712803
	0.4	0.05986398	0.05984741	0.05984633	0.06984626
	0.6	0.04719460	0.04718395	0.04718288	0.04718280
	0.8	0.02811236	0.02809929	0.02809836	0.02809829
	1.0	0	0	0	0
10	0	0.01393546	0.01392346	0.01392229	0.01392221
	0.2	0.01359641	0.01358865	0.01358778	0.01358769
	0.4	0.01256880	0.01254040	0.01253856	0.01253849
	0.6	0.01051380	0.01060370	0.01060192	0.01060180
	0.8	0.00729842	0.00728008	0.00727871	0.00727860
	1.0	0	0	0	0

$\dfrac{T_{2i-2}(x)}{\left(1-x^2\right)^{1/2}}$ for $i = 1, 2, ...,$ and then integrating with respect to the variable x, in the range $(-1,1)$, we arrive at the following system of linear equations.

$$\sum_{j=1}^{\infty} b_j \left[\pi \, \delta_{1j} - \frac{16}{\lambda \pi}(j-1)\left(\sum_{l=0}^{j-2} \frac{1}{2l+1} \right) \right] = \frac{1}{\lambda \pi} + \pi \quad \text{for } i = 1, \tag{6.3.22}$$

$$\sum_{j=1}^{\infty} b_j \left[\frac{\pi}{2} \delta_{ij} - \frac{16}{\lambda\pi}(j-1)\left(\sum_{l=0}^{j-2} \frac{2l+1}{(2l+2i-1)(2l-2i+3)} \right) \right] \tag{6.3.23}$$

$$= \frac{1}{\lambda\pi} \frac{1}{1-4(i-1)^2} \quad \text{for } i = 2,3,...,$$

where δ_{ij} denotes the Kronecker delta sybol, after using the following results

(i) $$\int_{-1}^{1} \frac{T_i(t)T_j(t)}{\left(1-t^2\right)^{1/2}} dt = \begin{cases} \pi, & \text{for } i = j = 0 \\ \dfrac{\pi}{2}, & \text{for } i = j \neq 0, \\ 0, & \text{for } i \neq j; \end{cases} \tag{6.3.24}$$

(ii) $$\int_{-1}^{1} \frac{U_{2j-3}(t)}{\left(1-t^2\right)^{1/2}(t-x)} dt = 2\pi \left\{ U_{2j-4}(x) + U_{2j-6}(x) + \cdots + U_2(x) + U_0(x) \right\}, \ -1<x<1; \tag{6.3.25}$$

(iii) $$\int_{-1}^{1} U_{2l}(t)T_{2i-2}(t)dt = \frac{4l+2}{(2l+2i-1)(2l-2i+3)}, \quad i,l=0,1,2,..., \tag{6.3.26}$$

(iv) $$\int_{-1}^{1} T_{2i-2}(t)dt = \frac{2}{1-4(i-1)^2}. \tag{6.3.27}$$

The above infinite system of linear equations are solved for the b_j's, by first truncating the infinite series in the relations (6.3.19) to (6.3.23), at an integral value N, say. In this case we consider the form

$$\varphi(x) \approx \varphi_N(x) = \sum_{j=1}^{N} b_j^{(N)} \left(T_{2j-2}(x) - T_0(x) \right), \quad -1 \leq x \leq 1, \tag{6.3.28}$$

as a possible approximation to the solution of the singular integro-differential equation. The system of equation (6.3.22) and (6.3.23) then takes the form

$$\sum_{j=1}^{N} b_j^{(N)} \left[\pi \delta_{1j} - \frac{16}{\lambda\pi}(j-1)\left(\sum_{l=0}^{j-2} \frac{1}{2l+1} \right) \right] = \frac{1}{\lambda\pi} + \pi \text{ for } i = 1 \tag{6.3.29}$$

$$\sum_{j=1}^{N} b_j^{(N)} \left[\frac{\pi}{2} \delta_{ij} - \frac{16}{\lambda\pi}(j-1)\left(\sum_{l=0}^{j-2} \frac{2l+1}{(2l+2i-1)(2l-2i+3)} \right) \right]$$

$$= \frac{1}{\lambda\pi} \frac{1}{1-4(i-1)^2} \quad \text{for } i = 2,3,...,N. \tag{6.3.30}$$

The system of linear equations (6.3.29) and (6.3.30) has been solved for different values of N ($N = 21, 41, 81, 161$) and the values of $\varphi_N(x)$, as given by the relation (6.3.28), are tabulated in Table 3 (taken from Chakrabarti and Hamsapriye(1999)), upto seventh decimal place accuracy.

Table 3
$\varphi_N(x)$ (Method 3)

λ	x_i / N	10	20	40	80
0.1	0	0.1157803	0.1157838	0.1157847	0.1157850
	0.2	0.1112375	0.1112408	0.1112417	0.1112419
	0.4	0.0975728	0.0975758	0.0975766	0.0975768
	0.6	0.0746703	0.0746729	0.0746735	0.0974673
	0.8	0.0423029	0.0423046	0.0423050	0.0423051
	1.0	0	0	0	0
1	0	0.0694869	0.0695024	0.0695064	0.0695074
	0.2	0.0671076	0.0671228	0.0671267	0.0671276
	0.4	0.0598272	0.0598414	0.0598450	0.0698459
	0.6	0.0471663	0.0471786	0.0471817	0.0471825
	0.8	0.0280865	0.0280865	0.0280975	0.0280981
	1.0	0	0	0	0
10	0	0.0139003	0.0139175	0.0139210	0.0139219
	0.2	0.0135690	0.0135831	0.0135565	0.0135873
	0.4	0.0125202	0.0125339	0.0125373	0.0125382
	0.6	0.0105854	0.0105975	0.0106007	0.0106015
	0.8	0.0072632	0.0072748	0.0072776	0.0072784
	1.0	0	0	0	0

Before this discussion is closed, a study on the convergence aspects for the Method 1 has been taken up. For this, the singular integro-differential equation (6.3.1), recast into the form (6.3.3) and expressible in the form (6.3.5), is considered here.

Convergence of Method 1

In the Method 1 (i.e. the collocation method), we have worked with an approximation to $\varphi(x)$ of the form (cf. the relation (6.3.9a))

$$\varphi(x) \approx \varphi_N(x) = \sum_{j=0}^{N} {}^{''} c_j^{(N)} \, T_j(x), \quad -1 < x < 1 \qquad (6.3.31)$$

where $c_j^{(N)}$'s are the constants which can be determined as described earlier. By suitably adjusting the coefficients in this approximation, we can rewrite $\varphi_N(x)$ in a more convenient form as

$$\varphi_N(x) = \sum_{j=0}^{N} \alpha_j^{(N)} \, \psi_j(x), \quad -1 < x < 1 \qquad (6.3.32)$$

where $\alpha_j^{(N)}$'s are certain constants and $\psi_j(x)$'s are defined by

$$\psi_j(x) = \left(\frac{2}{\pi}\right)^{1/2} T_j(x), \quad j = 1, 2, \ldots \qquad (6.3.33)$$

such that

$$\int_{-1}^{1} \rho(x) \, \left\{\psi_j(x)\right\}^2 \, dx = 1$$

where

$$\rho(x) = \left(1 - x^2\right)^{-1/2}.$$

In the convergence of the Method 1, it is required to prove that $\varphi_N(t)$ tends to $\varphi(x)$, in some sense, as $N \to \infty$. We shall show below that, in certain normed linear space of functions,

$$\|\varphi_N - \varphi\|_\rho \to 0 \text{ as } N \to \infty \qquad (6.3.34)$$

where $\|\cdot\|_\rho$ denotes the norm, as will be explained in the sequel.

Firstly, we show that the equation (6.3.3) or (6.3.5) (i.e. the singular integro-differential equation), is an operator equation between two Hilbert spaces. Let $L(\rho)$ denote the space of all real, square integrable functions with respect to the weight function $\rho(x) = \left(1 - x^2\right)^{-1/2}$.

The inner product of any two functions $u(x)$ and $v(x)$ belonging to $L(\rho)$ and the norm are defined by

$$< u, v >_\rho = \int_{-1}^{1} \frac{u(x) \, v(x)}{\left(1 - x^2\right)^{1/2}} dx \text{ and } \|u\|_\rho = \left\{< u, u >_\rho\right\}^{1/2} \qquad (6.3.35)$$

respectively. It is well known that $\{\psi_j\}_{j=0}^{\infty}$ as defined by the relation (6.3.33) forms a complete orthonormal system in the space $L(\rho)$.

Thus, for any $\varphi \in L(\rho)$, we can write

$$\varphi(x) = \sum_{j=0}^{\infty} <\varphi, \psi_j>_\rho \; \psi_j(x),$$

$$\|\varphi\|_\rho = \left[\sum_{j=0}^{\infty} \left\{ <\varphi, \psi_j>_\rho \right\}^2 \right]^{1/2} < \infty, \qquad (6.3.36)$$

the norm being obtained by using Parseval's relation

We consider the subspace $L_1(\rho)$ of $L(\rho)$, which consists of all those elements u, such that

$$\sum_{j=0}^{\infty} j^2 \left\{ <u, \psi_j>_\rho \right\}^2 < \infty \qquad (6.3.37)$$

for the reason, which is clear from the system of linear equations (6.3.12) and (6.3.13), so obtained in the Method 1, for solving the unknowns $c_j^{(N)}$.

By defining $v_j = \psi_j / j$ for $j \geq 1$, we obtain a complete orthonormal system in the space $L_1(\rho)$. The subspace $L_1(\rho)$ can be made into a Hilbert space by defining the inner product of any two functions $u(x)$, $v(x) \in L_1(\rho)$ and the norm to be

$$<u, v>_1 = \sum_{j=1}^{\infty} j^2 <u, \psi_j>_\rho <v, \psi_j>_\rho \qquad (6.3.38)$$

and

$$\|u\|_1 = \left[\sum_{j=1}^{\infty} j^2 \left\{ <u, \psi_j>_\rho \right\}^2 \right]^{1/2} < \infty \qquad (6.3.39)$$

respectively. It can be verified that $\|v_j\|_1 = 1, \; j = 1, 2, ...,$

It can be verified that D is a Hilbert-Schmidt operator from $L_1(\rho)$ into $L(\rho)$ and hence compact (cf. Golberg (1985)). Further, it can be shown that $\|Cu\|_\rho = \|u\|_1$.

Thus we see that $\left(D + \dfrac{2}{\lambda\pi} C \right)$ is a mapping from the space $L_1(\rho)$ into the space $L(\rho)$. In fact $\left(D + \dfrac{2}{\lambda\pi} C \right)$ is an operator from a much larger space into $L(\rho)$.

It can further be verified that $C^{-1} : L(\rho) \to L_1(\rho)$ exists and is given by

$$C^{-1}u = \sum_{j=1}^{\infty} <u,\psi_j>_\rho \frac{\psi_j}{j}, \quad u \in L(\rho) \qquad (6.3.40)$$

since $C\psi_j = j\psi_j$.

Now the operator D is compact and C^{-1} exists. Thus $\left(D + \frac{2}{\lambda\pi}C \right)$ has a bounded inverse (cf. Golberg (1985)) if and only if the null space of the operator $\left(D + \frac{2}{\lambda\pi}C \right)$ contains only the zero element. That is, the corresponding homogeneous equation of (6.3.3) has zero as its only solution. This is now assumed for some values of λ (see later). Therefore, we can find a positive real constant K, such that for the norm of the inverse operator $\left(D + \frac{2}{\lambda\pi}C \right)^{-1}$ (for $\lambda > 0$, from $L(\rho)$ to $L_1(\rho)$), there holds the estimation

$$\left\| \left(D + \frac{2}{\lambda\pi}C \right)^{-1} \right\|_\rho \le K. \qquad (6.3.41)$$

To prove that $\varphi_N(x)$ converges to $\varphi(x)$, we have to show that $\varphi_N(x)$ satisfies an analogous operator equation as $\varphi(x)$. For this purpose, we consider the interpolatory projection map (cf. Atkinson (1997)) $P_N : L(\rho) \to L(\rho)$, that maps $\varphi(x) \in C^0\left([-1,1]\right)$ onto the unique polynomial which interpolates $\varphi(x)$ at certain $(N+1)$ distinct points x_j, $j = 0,1,...,N$.

Replacing φ by φ_N in the relation (6.3.5) and using the usual collocation procedure, we arrive at

$$\left(D\varphi_N + \frac{2}{\lambda\pi}C\varphi_N - F \right)(x_j) = 0, \quad x_j = -\cos\frac{j\pi}{N}, \ j = 0,1,...,N. \quad (6.3.42)$$

Thus φ_N satisfies the relation (cf. Atkinson (1997))

$$P_N\left(D\varphi_N + \frac{2}{\lambda\pi}C\varphi_N - F \right) = 0. \qquad (6.3.43)$$

Since $P_N C\varphi_N = C\varphi_N$, we can write the relation (6.3.43) as

$$P_N D\varphi_N + \frac{2}{\lambda\pi} C\varphi_N = P_N F. \qquad (6.3.44)$$

Further, since C^{-1} exists and D is compact, there exists $N \geq N_0$ where N_0 is a large positive integer such that $\left(P_N D + \dfrac{2}{\lambda\pi} C \right)^{-1}$ exists (for $\lambda > 0$), and $\left\| (P_N D + \frac{2}{\lambda\pi} C)^{-1} \right\|_\rho \leq K'$ where K' is a constant (cf. Golberg (1985)).

By using the invertibility of the operator $\left(P_N D + \dfrac{2}{\lambda\pi} C \right)$, it can be proved that

$$\varphi - \varphi_N = \frac{2}{\lambda\pi} \left(P_N D + \frac{2}{\lambda\pi} C \right)^{-1} \left(C\varphi - P_N C\varphi \right) \text{ for } N \geq N_0. \ (6.3.45)$$

Thus, we obtain

$$\left\| \varphi - \varphi_N \right\|_\rho \leq \frac{2}{\lambda\pi} K' \left\| C\varphi - P_N C\varphi \right\|_\rho \text{ for } N \geq N_0. \qquad (6.3.46)$$

Now we have

$$C\varphi - P_N C\varphi = (C\varphi - p) - P_N (C\varphi - p) \qquad (6.3.47)$$

for all polynomials p of degree $\leq N$, since then $P_N p = p$, resulting in the relation

$$\left\| C\varphi - P_N C\varphi \right\|_\rho \leq \left\{ 1 + \left\| P_N \right\|_{C^o \to L(\rho)} \right\} \left\| C\varphi - p \right\|_\rho. \qquad (6.3.48)$$

Now, from the operator equation

$$\frac{2}{\lambda\pi} C\varphi = F - D\varphi$$

it can be argued that $C\varphi$ is at least once differentiable. This is derived from the fact that $\varphi(x) \in C^1 \left([-1,1] \right)$. A proof of this is given below. Using the above observations and assuming that p is a polynomial (of degree $\leq N$) of best uniform approximation to $C\varphi$ and using the fact that P_N is bounded, we obtain

$$\left\|\phi - \phi_N\right\|_\rho \le \frac{2}{\lambda \pi} K' \left\{1 + \left\|P_N\right\|_{C^0 \to L(\rho)}\right\} \left\|C\phi - p\right\|_\rho$$

$$\le \frac{2}{\lambda \pi} K' \left\{1 + \left\|P_N\right\|_{C^0 \to L(\rho)}\right\} \left\|C\phi - p\right\|_\infty \int_{-1}^{1} \rho(x)\, dx$$

(6.3.49)

$$= \frac{2}{\lambda} K' \left\{1 + \left\|P_N\right\|_{C^0 \to L(\rho)}\right\} \left\|C\phi - p\right\|_\infty$$

$$= 0\left(N^{-r}\right) \text{ for } N \ge N_0, \ r \ge 1,$$

after applying Jackson's theorem (cf. Baker (1977), Atkinson (1997)), so that

$$\left\|\phi - \phi_N\right\|_\rho \to 0 \text{ as } N \to \infty.$$

(6.3.50)

This proves the convergence of the Method 1.

Two important results

We prove here two important results, which have been used in the above analysis.

Result 1: $\phi(x) \in C^1\left([-1,1]\right)$ where $C^1\left([-1,1]\right)$ denotes the space of once continuously differentiable functions on the interval $[-1,1]$.

Proof: We know that

$$\phi \in L_1(\rho) \subset L(\rho) \text{ and } \phi(\pm 1) = 0.$$

These imply that

$$(T\phi)(x) = \fint_{-1}^{1} \frac{\phi(t)}{x-t}\, dt \in L(\rho)$$

(6.3.51)

since the singular integral operator is bounded (cf. Wolfersdorf (1983)). Thus from the integro-differential equation (6.3.1) we immediately see that $\dfrac{\partial \phi}{\partial x} \in L(\rho)$. Now,

$$\phi(x) = \phi(-1) + \int_{-1}^{x} \frac{d\phi}{dt}\, dt = \int_{-1}^{x} \frac{d\phi}{dt}\, dt.$$

(6.3.52)

Thus, we can estimate the quantity $|\varphi(x_2)-\varphi(x_1)|$, for any $x_1,x_2 \in [-1,1]$ to be

$$|\varphi(x_2)-\varphi(x_1)| = \left| \int_{x_1}^{x_2} \frac{d\varphi}{dt} \, dt \right|$$

$$\leq \left[\left\{ \int_{x_1}^{x_2} \left(\frac{d\varphi}{dt} \right)^2 dt \right\} \left\{ \int_{x_1}^{x_2} 1^2 \, dt \right\} \right]^{1/2} \quad (6.3.53)$$

$$\leq K |x_2 - x_1|^{1/2},$$

where K is defined to be

$$K = \left[\int_{x_1}^{x_2} \frac{1}{\left(1-t^2\right)^{1/2}} \left(\frac{d\varphi}{dt} \right) dt \right]^{1/2}. \quad (6.3.54)$$

The inequality (6.3.53) shows that $\varphi(x)$ is Hölder continuous, with exponent $1/2$ in $[-1,1]$. Again, since $\varphi(\pm 1) = 0$, the CPV integral $\int_{-1}^{1} \frac{\varphi(t)}{x-t} dt \, (-1 < x < 1)$ is Hölder continuous with exponent $1/2$ (cf. Muskhelishvili (1953)). Thus using the equation (6.3.1) as well as the above observations, we find that $\frac{d\varphi}{dx}$ is also Hölder continuous and hence continuous on $[-1,1]$.

Result 2: The homogeneous integro-differential equation (6.3.1) with $\varphi(\pm 1) = 0$ has only one solution $\varphi \equiv 0$ for $\lambda > 0$.

Proof: We consider the homogeneous integro-differential equation

$$2\frac{d\varphi}{dx} - \lambda \int_{-1}^{1} \frac{\varphi(t)}{x-t} dt = 0, \quad -1 < x < 1 \quad (6.3.55)$$

with $\lambda > 0$. Multiplying both sides by $\frac{d\varphi}{dx}$ and integrating with respect to x over $[-1,1]$, we obtain

$$2 \int_{-1}^{1} \left(\frac{d\varphi}{dx} \right)^2 dx + \lambda\pi < S\varphi, \varphi > \, = 0 \quad (6.3.56)$$

where $< \, , \, >$ is the usual inner product in the space L and that

$$(S\varphi)(t) = -\frac{1}{\pi} \int_{-1}^{1} \frac{\varphi'(x)}{x-t} \, dx, \quad -1 < t < 1. \tag{6.3.57}$$

The second term in the relation (6.3.56) satisfies the inequality (cf. Wolfersdrof (1983))

$$(S\varphi,\varphi) \geq \int_{-1}^{1} \frac{\{\varphi(x)\}^2}{\left(1-x^2\right)^{1/2}} \, dx \geq 0. \tag{6.3.58}$$

Thus it is clear that each term in the relation (6.3.56) has to be equal to zero, and in particular, $\dfrac{d\varphi}{dx} = 0$ which implies that $\varphi(x) = 0$ since $\varphi(\pm 1) = 0$.

In the next two sections we highlight two simpler approximate methods of solution of the integro-differential equation (6.3.1), which appear to be simple and straightforward in comparison with the Galerkin methods discussed above. The convergences of the methods are also discussed.

6.4 APPROXIMATE METHOD BASED ON POLYNOMIAL APPROXIMATION

The Cauchy type singular integral equation

$$2\frac{d\varphi}{dx} - \lambda \int_{-1}^{1} \frac{\varphi(t)}{x-t} \, dt = f(x), \quad -1 < x < 1, \, \lambda > 0, \tag{6.4.1}$$

with $\varphi(\pm 1) = 0$ has been considered for numerical solution employing a simple method based on polynomial approximation of the function $\varphi(x) \, (-1 < x < 1)$ by Mandal and Bera (2007). This simple method is discussed here. Since the unknown function $\varphi(x)$ satisfies the end condition $\varphi(\pm 1) = 0$, it can be represented in the form

$$\varphi(x) = \left(1-x^2\right)^{1/2} \psi(x), \quad -1 \leq x \leq 1 \tag{6.4.2}$$

where $\psi(x)$ is a well behaved unknown function of x in (-1.1). We approximate the unknown function $\psi(x)$ by means of a polynomial of degree n, as given by

$$\psi(x) = \sum_{j=0}^{n} a_j x^i \qquad (6.4.3)$$

where a_j's $(j = 0,1,...,n)$ are unknown constants, then the equation (6.4.1) reduces to

$$\sum_{j=0}^{n} a_j \alpha_j(x) = g(x), \quad -1 < x < 1 \qquad (6.4.4)$$

where

$$\alpha_0(x) = -\frac{x}{\left(1-x^2\right)^{1/2}} - \frac{\lambda\pi}{2}x,$$

$$\alpha_j(x) = \frac{jx^{j-1} - (j+1)x^{j+1}}{\left(1-x^2\right)^{1/2}} + \frac{\lambda}{2}\left\{-\pi x^{j+1} + \sum_{k=0}^{j-1} \frac{1+(-1)^k}{4} \frac{\Gamma\left(\frac{1}{2}\right)\Gamma\left(\frac{k+1}{2}\right)}{\Gamma\left(\frac{k}{2}+2\right)} x^{j-k-1}\right\}, \quad j=1,2,...,n.$$

and

$$g(x) = \frac{1}{2}f(x). \qquad (6.4.5)$$

The unknown constants $a_j(j = 0,1,...,n)$ are now obtained by putting $x = x_l (l = 0,1,...,n)$ in (6.4.4) where x_l's are distinct and $-1 < x_l < 1$. Thus we obtain a system of $(n+1)$ linear equations, given by

$$\sum_{j=0}^{n} a_j \alpha_{jl} = g_l, \ l = 0,1,...,n, \qquad (6.4.6)$$

where

$$\alpha_{jl} = \alpha_j(x_l), \ g_l = g(x_l). \qquad (6.4.7)$$

The method is now illustrated for $f(x) = -x/2$ as considered in section 6.3. However, other forms of $f(x)$ can also be considered. For the linear system (6.4.6), we choose $n = 10$, and the collocation points as $x_0 = -0.924$, $x_1 = -0.665$, $x_3 = -0.408$, $x_4 = -0.023$, $x_{10} = 0.961$ and $\lambda = 1$. Since $f(x) = -x/2$, we have $g(x) = -x/4$. The system of linear equations (6.4.6)now produces $a_0 = 0.070$, $a_1 = 0.000$, $a_2 = -0.024$, $a_3 = 0.000$, $a_4 = -0.004$, $a_5 = -0.003$, $a_6 = -0.035$ $a_7 = 0.011$, $a_8 = 0.061$, $a_9 = -0.011$, $a_{10} = -0.052$. Using these coefficients, the values of $\varphi(x)$ at $x = 0.2k$, $k = 0,1,...5$

are presented in Table 1 below. The values of $\varphi(x)$ given in Table 1 of section 6.3 at these points (for $\lambda = 1$) are also given (correct upto three decimal places) for comparison.   It is obvious that the results obtained by the present method compares favourably with the results obtained in section 6.3.

Table 1

x_k		0.0	0.2	0.4	0.6	0.8	1.0
$\varphi(x_k)$	Present Method	0.070	0.068	0.061	0.048	0.029	0
	Method Of Sec. 6.3	0.069	0.067	0.060	0.047	0.028	0

The present choice of collocation points which are not equispaced, helps in casting the original problem of integro-differential equation (6.4.1) with $\lambda = 1$, $f(x) = -\dfrac{x}{2}$ into a system of algebraic equations where appearance of ill-conditioned matrices have been avoided altogether.

Convergence of the method

Substitution of $\varphi(x)$ in terms of $\psi(x)$ given by (6.4.2) into the equation (6.4.1) produces an equation for $\psi(x)$. Then $\psi(x)$ satisfies the equation

$$\left(D - \frac{\lambda\pi}{2}C\right)\psi(x) = g(x), \quad -1 < x < 1 \tag{6.4.8}$$

where C, D respectively denote the operators defined by

$$(Cu)(x) = \frac{1}{\pi}\int_{-1}^{1} \frac{\left(1-t^2\right)^{1/2}}{x-t}\, u(t)\, dt, \quad -1 < x < 1 \tag{6.4.9}$$

and

$$(Du)(x) = \left(1-x^2\right)^{1/2}\frac{du}{dx} - \frac{x}{\left(1-x^2\right)^{1/2}}\, u(x), \quad -1 < x < 1. \tag{6.4.10}$$

Let $T_n(x) = \cos n\theta$ with $x = \cos\theta$, be the Chebyshev polynomial of the first kind. Then

$$(CU_n)(x) = -T_{n+1}(x), \quad n \geq 0 \tag{6.4.11}$$

where $U_n(x)$ is a Chebyshev polynomial of the second kind.

This shows that the operator C can be extended as a bounded linear operator from $L_1(\mu)$ to $L(\mu)$, where $L(\mu)$ is the subspace of functions square integrable with respect to $\mu(x) = (1-x^2)^{-1/2}$ and $L_1(\mu)$ is the subspace of functions $u \in L(\mu)$ satisfying

$$\|u\|_1^2 = \sum_{k=0}^{\infty} (k+1)^2 <u, \psi_{k+1}>_\mu^2 \ < \infty \tag{6.4.12}$$

where

$$\psi_k = \left(\frac{2}{\pi}\right)^{1/2} T_k, \tag{6.4.13}$$

and

$$<u, v>_\mu = \int_{-1}^{1} \mu(t)\, u(t)\, v(t)\, dt. \tag{6.4.14}$$

Again,

$$(DU_n)(x) = -\frac{n+1}{(1-x^2)^{1/2}} T_{n+1}(x), \quad n \geq 0. \tag{6.4.15}$$

This shows that D can be extended as a linear operator from $L_1(\mu)$ to $L(\mu)$. Assuming $g \in L(\mu)$, we find that the equation (6.4.8) possesses a unique solution $\psi \in L_1(\mu)$ for each $g \in L(\mu)$.

If we use the polynomial approximation (6.4.3) for ψ, then

$$\psi(x) \approx p_n(x) = \sum_{j=0}^{n} a_j x^j. \tag{6.4.16}$$

Since x^j $(j = 0, 1, ..., n)$ can be expressed in terms of Chebyshev polynomials of first kind $T_m(x)$ $(m = 0, 1, ..., j)$ as (cf. Snyder (1966))

$$x^j = \frac{1}{2^{j-1}} \sum_{k=0}^{[j/2]} \binom{j}{k} T_{j-2k}(x) \tag{6.4.17}$$

where $[j/2]$ denotes the greatest integer $\leq j/2$, we can express $p_n(x)$ given by (6.4.16) as

$$p_n(x) = \sum_{i=0}^{n} b_i \, T_i(x), \tag{6.4.18}$$

where the coefficients b_i $(i = 0,1,...,n)$ can be expressed in terms of a_j $(j = 0,1,...,i)$ and vice-versa. The right side of (6.4.18) is now denoted by

$$u_n(x) = \sum_{k=0}^{n} c_k \, \psi_k(x) \tag{6.4.19}$$

where

$$c_k = \left(\frac{\pi}{2}\right)^{1/2} b_k.$$

To determine an error estimate in replacing ψ by p_n, we note that

$$\|\psi - p_n\|_1 = \|\psi - u_n\|_1. \tag{6.4.20}$$

Following the reasoning given in Golberg and Chen (p.309, 1997), it can be shown that

$$\|\psi - u_n\|_1 < \frac{c}{n^r} \tag{6.4.21}$$

where c is a constant and r is such that $g \in C^r[-1,1]$. In the above computation $g(x) = -\dfrac{x}{4}$ and thus is a C^∞ function. Hence r in (6.4.21) can be chosen very large so that the error becomes negligible as n increases, and the convergence is quite fast. This is also reflected in the above numerical results.

6.5 APPROXIMATE METHOD BASED ON BERNSTEIN POLYNOMIAL BASIS

The integro-differential equation

$$2\frac{d\varphi}{dx} - \lambda \int_{-1}^{1} \frac{\varphi(t)}{x-t} dt = f(x), \quad -1 \leq x \leq 1 \tag{6.5.1}$$

with $\varphi(\pm 1) = 0$, has also been solved numerically employing the method of polynomial approximation in the Bernstein polynomial basis by Bhattacharya and Mandal (2008). This method is briefly discussed in this section.

Bernstein polynomials

The Bernstein polynomials of degree n over the interval $[0,1]$ are defined by

$$B_{i,n}(x) = \binom{n}{i} x^i (1-x)^{n-i}, i = 0,1,...,n. \tag{6.5.2}$$

Bernstein first used these polynomials in the proof of Wiestrass theorem. By a linear transformation, any interval $[a,b]$ can be changed to $[0,1]$, so that without any loss of generality one can consider a function to be defined on $[0,1]$. The corresponding approximation of the function $f(x)$ on the Bernstein polynomial basis in the interval $[0,1]$ is given by

$$B_n^f(x) = \sum_{i=0}^{n} a_i \binom{n}{i} x^i (1-x)^{n-i} \tag{6.5.3}$$

where a_i is defined as $a_i = f\left(\dfrac{i}{n}\right)$. (6.5.3) is a Bernstein polynomial approximation of the function $f(x)$ and it can be proved that for a function $f(x)$ bounded on $[0,1]$,

$$\lim_{n \to \infty} B_n^f(x) = f(x) \tag{6.5.4}$$

holds at each point of continuity x of $f(x)$ and that the relation holds uniformly on $[0,1]$ if $f(x)$ is uniformly continuous on the interval (cf. Lorenz (1953)). In fact by making suitable changes in the definition of the coefficients a_i in the expression (6.5.3), this property can be extended to approximate any general function. The coefficients a_i in that case can be an arbitrary constant. Hence Bernstein polynomial basis provides a very good basis for approximation of unknown function satisfying an integral equation.

Bernstein polynomials of degree 1 are given by

$$B_{0,1}(x) = 1 - x, \quad B_{1,1}(x) = x;$$

Bernstein polynomials of degree 2 are given by

$$B_{0,2}(x) = (1-x)^2, \quad B_{1,2} = 2x(1-x), \quad B_{2,2} = x^2;$$

Bernstein polynomials of degree 3 are given by

$$B_{0,3}(x) = (1-x)^3, \quad B_{1,3}(x) = 3x(1-x)^2, \quad B_{2,3} = 3x^2(1-x), \quad B_{3,3}(x) = x^3;$$

and so on.

Properties of Bernstein polynomials

(a) Recursive relation:

$$B_{i,n}(x) = (1-x)B_{i,n-1}(x) + x_{i-1,n-1}(x).$$

(b) Derivatives

$$\frac{d}{dx}B_{i,n}(x) = B_{i-1,n-1}(x) - B_{i,n-1}(x).$$

(c) Positivity

$$B_{i,n}(x) \text{ is positive for } x \in [0,1].$$

(d) Symmetry

$$B_{i,n}(x) = B_{n-i,n}(x).$$

(e) Sum to unity

$$\sum_{i=0}^{n} B_{i,n}(x) = 1.$$

Approximate solution using Bernstein polynomial basis

The unknown function $\varphi(x)$ of (6.5.1) with $\varphi(\pm 1) = 0$, can be represented in the form

$$\varphi(x) = \left(1-x^2\right)^{1/2}\psi(x), \quad -1 \le x \le 1 \qquad (6.5.5)$$

where $\psi(x)$ is a well behaved function of $x \in [0,1]$. To find an approximate solution of (6.5.1), $\psi(t)$ is approximated using Bernstein polynomials in $[-1,1]$ as

$$\psi(t) \approx \sum_{i=0}^{n} a_i B_{i,n}(t) \qquad (6.5.6)$$

where now $B_{i,n}(x)$ $(i = 0,1,...,n)$ are defined on $[-1,1]$ as

$$B_{i,n}(x) = \binom{n}{i} \frac{(1+x)^i (1-x)^{n-i}}{2^n}, \quad i = 0,1,...,n \qquad (6.5.7)$$

and a_i $(i = 0,1,...,n)$ are unknown constants to be determined. Substituting (6.5.7) in (6.5.1) we get

$$\sum_{i=0}^{n} a_i \left[-\frac{2x}{(1-x^2)^{1/2}} B_{i,n}(x) + 2(1-x^2)^{1/2} B'_{i,n}(x) - \lambda \int_{-1}^{1} (1-t^2)^{1/2} \frac{B_{i,n}(t)}{x-t} dt \right] \qquad (6.5.8)$$

$$= f(x), \quad -1 \le x \le 1.$$

Multiplying both sides by $B_{j,n}(x)$ $(j = 0,1,...,n)$ and integrating from -1 to 1 we get a linear system given by

$$\sum_{i=0}^{n} a_i C_{ij} = b_j, \quad j = 0,1,...,n \qquad (6.5.9)$$

where

$$C_{ij} = -2 \int_{-1}^{1} \frac{x}{(1-x^2)^{1/2}} B_{i,n}(x) B_{j,n}(x) dx + 2 \int_{-1}^{1} (1-x^2)^{1/2} B'_{i,n}(x) B_{j,n}(x) dx$$

$$+ \lambda \int_{-1}^{1} \left\{ \int_{-1}^{1} (1-t^2)^{1/2} \frac{B_{i,n}(t)}{x-t} dt \right\} B_{j,n}(x) dx \qquad (6.5.10)$$

and

$$b_j = \int_{-1}^{1} f(x) B_{j,n}(x) dx. \qquad (6.5.11)$$

For $\lambda = 1$, we can write C_{ij} as

$$C_{ij} = D_{ij} + E_{ij} + F_{ij}$$

where

$$D_{ij} = -n \int_{-1}^{1} (1-x^2)^{1/2} B_{i,n}(x) B_{j-1,n-1}(x) dx$$

$$= -4j \binom{n}{i} \binom{n}{j} \frac{\Gamma\left(i+j+\frac{1}{2}\right) \Gamma\left(2n-i-j+\frac{3}{2}\right)}{\Gamma(2n+2)}, \, j = 1,2,...,n; \, i = 0,1,...,n, \qquad (6.5.12)$$

$$E_{ij} = n \int\limits_{-1}^{1} \left(1-x^2\right)^{1/2} B_{i,n}(x)\, B_{j,n-1}(x)\, dx$$

$$= 4(n-j) \binom{n}{i}\binom{n}{j} \frac{\Gamma\left(i+j+\frac{3}{2}\right)\Gamma\left(2n-i-j+\frac{1}{2}\right)}{\Gamma(2n+2)}, \tag{6.5.13}$$

$$j = 0,1,...,(n-1); \ i = 0,1,...,n$$

and

$$F_{ij} = \int\limits_{-1}^{1} A_i(x)\, B_{j,n}(x)\, dx$$

where

$$A_i(x) = \frac{1}{2^n}\binom{n}{i}\sum_{k=0}^{n} d_k^{i,n}\left[-\pi\, x^{k+1} + \sum_{m=0}^{k-1} \frac{1+(-1)^m}{4}\frac{\Gamma\left(\frac{m+1}{2}\right)\Gamma\left(\frac{1}{2}\right)}{\Gamma\left(\frac{m}{2}+2\right)} x^{k-m-1}\right], \tag{6.5.14}$$

$$i = 0,1,...,n$$

so that

$$F_{ij} = \frac{1}{2^{2n}}\binom{n}{i}\binom{n}{j}\sum_{k=0}^{n}\sum_{r=0}^{n} d_k^{i,n}\, d_r^{j,n}\left[-\pi\,\frac{1-(-1)^{k+r}}{k+r+2}\right.$$

$$\left. + \sum_{m=0}^{k-1}\frac{1-(-1)^{k+r-m}}{k+r-m}\frac{1+(-1)^m}{4}\frac{\Gamma\left(\frac{m+1}{2}\right)\Gamma\left(\frac{1}{2}\right)}{\Gamma\left(\frac{m}{2}+2\right)}\right], \ j = 0,1,...n; \ i = 0,1,...n \tag{6.5.15}$$

with

$$d_k^{i,n} = \sum_{s} (-1)^{k-s}\binom{i}{s}\binom{n-i}{k-s}, \ k = 0,1,...n; \ i = 0,1,...,n, \tag{6.5.16}$$

the summation over s being taken as follows:
for $i < n-i$ (i) $s = 0$ to k for $k \le i$,
(ii) $s = 0$ to i for $i < k \le n-i$, (iii) $s = k-(n-i)$ to $n-i$
for $n-i < k \le n$, while for
$i = n-i$ (n being an even integer) (i) $s = 0$ to k for $k \le i$,
(ii) $s = k-i$ to i for $i < k \le n$; for $i > n-i$, i and $n-i$ above
are to be interchanged. Also we find that for the choice of $f(x) = -\dfrac{x}{2}$,

$$b_j = \frac{1}{2^n}\binom{n}{j}(n-2j)\frac{\Gamma(n-j+1)\,\Gamma(j+1)}{\Gamma(n+3)}, \ j = 0,1,...n. \tag{6.5.17}$$

The system (6.5.9) is solved for unknowns a_i $(i = 0,1,...,n)$ by standard numerical method and numerical values of $\varphi(x)$ for different values of x are obtained approximately.

In our numerical calculations we have chosen $n = 7,10,13$ and $a_0, a_1,..., a_n$ are obtained numerically. Using these coefficients the values of $\varphi(x)$ at $x = (0.2)k$, $k = 0,1,...,5$ are presented in Table 1 below. For a comparison between the present method and that of the method used in Frankel (1998), values of $\varphi(x)$ at these points obtained by Frankel (1995) are also given. It is obvious that the result compares favourably with the results of Frankel and also obtained in Table 1 of section 6.3, and in Table 1 section 6.4.

Table 1

x	0	0.2	0.4	0.6	0.8	1.0
$\varphi(x)$ $n=7$	0.06973	0.06711	0.05964	0.04736	0.02811	0
(Present $n=10$	0.06948	0.06714	0.05988	0.04711	0.02821	0
method) $n=13$	0.06950	0.06717	0.05981	0.04723	0.02805	0
$\varphi(x)$ Frankel's method	0.06950	0.06712	0.05984	0.04718	0.02891	0

Convergence of the method can be proved as in section 6.4. In place of the relation (6.4.17), we have the relation (cf. Snyder (1969))

$$B_{i,n}(x) = \binom{n}{i} \frac{1}{2^n} \sum_{s=0}^{n} d_s^{i,n} \frac{1}{2^s} \sum_{m=0}^{\left[\frac{s}{2}\right]} \left\{ \binom{s}{m} - \binom{s}{m+1} \right\} U_{s-2m}(x). \quad (6.5.14)$$

where $d_s^{i,n}$ is given in (6.5.16).

Although the numerical computations have been carried out for $f(x) = -x/2$, the method can be applied for other forms of $f(x)$.

Chapter 7

Galerkin Method and its Application

The term Galerkin method has already been intoduced in section 6.3 without any explanation. In this chapter the underlying mathematical idea behind Galerkin method of determining approximate solution to a general operator equation $L f = l$ along with approximation of the inner product $[f,l]$, is explained where L is a linear operator. This will be applied to solve approximately singular integral equations with Cauchy type kernels. Also, application of the method to a number of water wave scattering problems involving thin vertical barriers arising in the linearised theory of water waves, as given in Mandal and Chakrabarti (1999), will be reviewed.

7.1 GALERKIN METHOD

Here we explain the general Galerkin method, which is also called a *projection method*, to solve any operator equation of the form

$$(L f)(x) = l(x), \quad x \in \mathcal{A} \qquad (7.1.1)$$

where L is a linear operator from a certain product space S to itself and $\mathcal{A}$ denotes a simply-connected domain in $\mathbb{R}^n$, in standard notations.

We first introduce the following important definitions.

Definition 1: A function $f(x)$ is said to solve the *operator equation* (7.1.1), if and only if,

$$[L f, \lambda] \equiv [\lambda, L f]^* = [\lambda, l]^* = [l, \lambda] \qquad (7.1.2)$$

for all $\lambda(x) \in S$, where the inner product $[u, v]$ is defined as

$$[u, v] = \int_{A} u(x) \, v^*(x) \, dx, \qquad (7.1.3)$$

the integral being understood in the ordinary Riemann sense and the stars (*) denote complex conjugates.

Definition 2: The solution function $f(x)$, as defined by the relations (7.1.2), associated with the equation (7.1.1), is said to be an *approximate solution* of the equation (7.1.1), if the relations (7.1.2) hold good, only *approximately*, i.e., if

$$[L \, f, \lambda] \equiv [\lambda, L \, f]^* \approx [\lambda, l]^* \qquad (7.1.4)$$

for all $\lambda(x) \in S.$

Note: In the above definition, the phrase, "a relation holding good approximately", means that the absolute value of the difference of the two sides of the relation is very small. For example, if we say that

$$a \approx b \qquad (7.1.5)$$

where a and b are *complex* numbers, then we must have

$$|a - b| < \varepsilon$$

where ε is a sufficiently small positive number.

As is clear from the above meaning of the *approximate solution* of the operator equation (7.1.1), it is found that the approximate relation (7.1.4) is to hold good instead of the actual equation (7.1.1). Also since the quantity $[L \, f, \lambda]$ represents the projection of the vector $L \, f$, onto the vector λ, in the usual sense whenever the inner products are interpreted geometrically, the approximate methods of solution of operator equations of the form (7.1.1), in the sense described above, are also called *projection methods.*

The method

In this section we are concerned with the Galerkin method of solving approximately the operator equation (7.1.1).   For this purpose we approximate $f(x)$ in the form

$$f(x) \approx F(x) = \sum_{j=1}^{N} a_j \, \varphi_j(x) \qquad (7.1.6)$$

where $\left\{\varphi_j(x)\right\}_{j=1}^{N}$ denotes a set, contained in S, of *independent* functions of $x \in \mathcal{A}$, which need not be an orthogonal set nor need be complete also, and a_j's $(j = 1, 2, ... N)$ are unknown constants to be determined.

Then taking the inner product of both sides of the operator equation (7.1.1), with f replaced by F, as given by the relation (7.1.6), we obtain

$$\sum_{j=1}^{N} a_j \; [L\varphi_j(x), \lambda(x)] \approx [l(x), \lambda(x)] \quad \text{for} \; \lambda \in S. \qquad (7.1.7)$$

Then by choosing $\lambda(x) \equiv \varphi_k(x)$, for some fixed positive integer k, such that $1 \leq k \leq N$, we obtain

$$\sum_{j=1}^{N} a_j \; [L \; \varphi_j(x), \; \varphi_k(x)] \approx [l(x), \; \varphi_k(x)], \; k = 1, 2, ..., N. \quad (7.1.8)$$

By treating then the approximate relations (7.1.8) as identities, as is customary in Galerkin methods, we obtain a system of N linear equations, for the N unknown constants $a_1, a_2, ..., a_N$, which can be solved easily by standard methods. The approximate solution of the operator equation (7.1.1), in the above described sense, is then given by the relation (7.1.6).

It is observed, from practical experience, that in many applications of integral equations, instead of getting the exact informations on the function $f(x)$, satisfying the actual operator equation (7.1.1), many informations of physical importance concerning a practical problem can be derived from the knowledge of the inner product $\mathcal{K} = [f, u]$, where $u(x)$ is a known function of $x \in \mathcal{A}$. It is then clear that the knowledge of the approximate function $F(x)$, as given by the relation (7.1.6), helps in determining the number $\mathcal{K}$, by means of the relation

$$\mathcal{K} = [f, u] \approx \sum_{j=1}^{N} a_j \left[\varphi_j, u \right], \qquad (7.1.9)$$

where the constants a_j's $(j = 1, 2, ..., N)$ are solutions of the system of linear equations (7.1.8).

Single-term approximation

In many problems in mathematical physics (cf. Jones p 269 (1966)) it is enough to use just a *single-term Galerkin approximation,* for which one can take $N = 1$, in the relation (7.1.6), and then find just one equation, in the linear system (7.1.8), giving

$$f(x) \approx F(x) = a_1 \, \varphi_1(x) \qquad (7.1.10)$$

with

$$a_1 = \frac{[l, \varphi_1]}{[L\varphi_1, \varphi_1]}. \qquad (7.1.11)$$

The approximate value of $\mathcal{K}$, in such circumstances, turn out to be given by the formula

$$\mathcal{K} = \frac{[l, \varphi_1][\varphi_1, u]}{[L\varphi_1, \varphi_1]}, \qquad (7.1.12)$$

Some important observations

We have

(i) $\quad [F, LF] \approx [F, l]$

(ii) $\quad [f, l] = [l, f] = [l, F] + [l, f - F]$

(iii) $\quad [l, f - F] = [l, f] - 2[l, F] + [l, F]$

$\qquad \approx [Lf, f] - 2[LF, F] + (F, LF)$ (by using (i))

(iv) $\quad [f - F, L(f - F)] = [f, Lf] - [f, LF] - [F, Lf) + [F, LF]$

$\qquad \approx [Lf, f] - 2[LF, F] + [F, LF]$

if $[Lh_1, h_2] = [h_1, Lh_2]$ for all $h_1, h_2 \in S$, i.e. if L is a *self-adjoint operator*.

Thus by using the results (iii) and (iv), we find that if L is a self-adjoint operator, we have that

$$[l, f - F] \approx [f - F, L(f - F)]. \qquad (7.1.13)$$

If, further we have either of the facts that (a) L is positive semi-definite, i.e., $[h, Lh] \geq 0$ for all $h \in S$ and (b) L is negative semi-definite, i.e., $[h, Lh] \leq 0$ for all $h \in S$, then we find from equation (7.1.13), that, either

$$(I)\quad [l,F] \le [l,f]\ \text{in case (a)},\qquad (7.1.14a)$$

or

$$(II)\quad [l,F] \ge [l,f]\ \text{in case (b)}.\qquad (7.1.14b)$$

The above two facts (I) and (II) imply that the quantity $[l,F]$, computed out of the approximate solution F of the operator equation (7.1.1), provides a lower bound for the actual quantity $[l,F]$ in the cases where L represents a *positive semi-definite* operator whereas $[l,F]$ provides an *upper bound* for the actual quantity $[l,f]$ in the cases where **L** represents a *negative semi-definite operator.*

Several problems of water wave scattering arising in the linearised theory of water waves, can be resolved approximately in the sense as described above, and bounds for certain useful quantities of the type $[l,f]$ for known l, can be determined approximately where one has to work simultaneously with a pair of operators in this class of problems. In many cases it has been observed that the two bounds, when computed numerically, agree up to two to three decimal places by employing the aforesaid single term approximations, and beyond six decimal places by employing multi-term approximations, so that their averages produce fairly accurate numerical estimates for the physical quantity $[l,f]$. This principle has been utilized successfully in many water wave scattering problems involving barriers.

In the next section we briefly describe the operator **L** arising in the study of a number of water wave scattering problems involving thin vertical barriers and give a list of exact solutions of approximate related problems, which are used in the single-term approximations.

7.2 USE OF SINGLE-TERM GALERKIN APPROXIMATION

In this section we describe a few water wave scattering problems for which single term Galerkin approximations have been utilized successfully to obtain accurate numerical estimates for the reflection and transmission coefficients.

The *oblique* water wave scattering problems involving a plane vertical thin barrier in deep water cannot be solved explicitly unlike the case when the incoming surface wave train is normally incident on the barrier. The surface-piercing vertical barrier was considered by Evans and Morris (1972a), who used the aforesaid single-term Galerkin approximation to obtain upper and lower bounds for the reflection and transmission coefficients. These bounds involve some definite integrals, which are rather straightforward to compute numerically. It has been found that these bounds, when computed numerically for various values of the different parameters, coincide up to two to three decimal places and as such their averages produce fairly accurate numerical estimates for the reflection and transmission coefficients.

In the course of mathematical analysis for the problem of oblique wave scattering by a thin vertical barrier present in deep water, the following integral equations arise (cf. Evans and Morris (1972a), Mandal and Das (1996))

$$\int_{\bar{L}} f(u) \, \mathcal{M}(y,u)du = e^{-Ky}, \ y \in \bar{L} \qquad (7.2.1a)$$

$$\int_{L} g(u) \, \mathcal{N}(y,u)du = e^{-Ky}, \ y \in L \qquad (7.2.1b)$$

where L denotes an interval whose length is equal to the length of the wetted portion of the vertical barrier, $\bar{L} = (0,\infty) - L$, K is a positive constant, $f(y)$ is proportional to the horizontal component of velocity in the gap above or below the barrier while $g(y)$ is proportional to the difference of velocity potential across the barrier so that $f(y)$ is required to have a square root singularity near an edge while $g(y)$ tends to zero as one approaches an edge, $\mathcal{M}(y,u)$ and $\mathcal{N}(y,u)$ are given by

$$\mathcal{M}(y,u) = \int_0^\infty \frac{(k \cos ky - K \sin ky)(k \cos ku - K \sin ku)}{\left(k^2 + v^2\right)^{1/2} \left(k^2 + K^2\right)} dk, \qquad (7.2.2a)$$

$$\mathcal{N}(y,u) = \lim_{\varepsilon \to 0+} \int_0^\infty \frac{\left(k^2 + v^2\right)^{1/2} (k \cos ky - K \sin ky)(k \cos ku - K \sin ku)}{k^2 + K^2} e^{-\varepsilon k} dk, y,u \in L \quad (7.2.2b)$$

with $v = K \sin \alpha$, α being the angle of incidence of a surface wave train incident upon the barrier from a large distance.

Thus, $\mathcal{M}(y,u)$ and $\mathcal{N}(y,u)$ are real symmetric functions of y and u, and $\mathcal{M}$ and $\mathcal{N}$ are positive semi-definite linear integral operators defined by

$$(\mathcal{M}f)(y) = \int_{\bar{L}} f(u) \, \mathcal{M}(y,u)du, \ y \in \bar{L},$$

$$(\mathcal{N}g)(y) = \int_{L} g(u) \, \mathcal{N}(y,u)du, y \in L.$$

Along with the integral equations (7.2.1a) and (7.2.1b) we have that

$$\int_L f(y)e^{-Ky}dy = C,$$ (7.2.3a)

$$\int_L g(y)e^{-Ky}dy = \frac{1}{\pi^2 K^2 C}$$ (7.2.3b)

where the real (unknown) constant C is related to the reflection and transmission coefficients (complex) R and T respectively by

$$C = \frac{(1-R)\cos\alpha}{i\pi R} = \frac{T\cos\alpha}{i\pi(1-T)}.$$ (7.2.4)

The integral equations (7.2.1a) and (7.2.1b) can be identified with the operator equation (7.1.1) while the relations (7.2.3a) and (7.2.3b) can be identified with the inner product (7.1.3), $\mathcal{A}$ denoting $\bar{L}$ or L and the inner product is simply the integral over $\mathcal{A}$.

It so happens that for *normal incidence* of the incoming wave train, the integral equations corresponding to (7.2.1a) and (7.2.1b) possess exact solutions. A single term Galerkin approximation to $f(y)$ in terms of the corresponding exact solution $f_0(y)$, say, for normal incidence of the wave train, provides a lower bound C_1 for C by noting the equality (7.2.3a) and using the inequality (7.1.14a) since $\mathcal{M}$ is a positive semi-definite linear operator. Similarly a single-term Galerkin approximation to $g(y)$ in terms of the corresponding exact solution $g_0(y)$, say, for normal incidence of the wave train, provides a lower bound $\frac{1}{C_2}$ for $\frac{1}{C}$ and hence an upper bound C_2 for C by noting the equality (7.2.3b) and using the inequality (7.1.14a) again since $\mathcal{N}$ is a positive semi-definite linear operator. Thus it is found that

$$C_1 \leq C \leq C_2.$$ (7.2.5)

Now, we have from the equation (7.2.4),

$$|R| = \frac{1}{\left(1+\pi^2 C^2 \sec^2\alpha\right)^{1/2}}, \quad |T| = \frac{\pi C\sec\alpha}{\left(1+\pi^2 C^2 \sec^2\alpha\right)^{1/2}}.$$ (7.2.6)

It is found that R_1 and R_2 obtained from (7.2.6) by using C_1 and C_2 respectively in place of C, provide upper and lower bounds for $|R|$. Similarly the bounds T_1 and T_2 for $|T|$ are obtained. Since $|R|^2 + |T|^2 = 1$ always, it is sufficient to consider $|R|$ only. At least for three configurations of the vertical barrier, viz. surface-piercing partially immersed barrier (Evans and Morris (1972a)), submerged vertical plate (Mandal and Das (1996)), and thin vertical semi-infinite barrier with a submerged gap (Das et al (1996)) it has been observed that R_1 and R_2 agree within two to three decimal places when computed numerically for any wave number and some particular values of different parameters. Thus averages of R_1 and R_2 produce fairly accurate estimates for $|R|$.

Four different configurations of the barrier are usually considered. For a surface-piercing partially immersed barrier, $L = (0, a)$, $\bar{L} = (a, \infty)$; for a submerged barrier extending infinitely downwards, $L = (a, \infty)$, $\bar{L} = (0, a)$; for a submerged plate $L = (a, b)$, $\bar{L} = (0, a) \cup (b, \infty)$ and for a vertical wall with a submerged gap, $L = (0, a) \cup (b, \infty)$, $\bar{L} = (a, b)$. We state below the functions $f_0(y)$ and $g_0(y)$ in terms of which the single-term Galerkin approximations of the integral equations (7.2.1a) and (7.2.1b) respectively are made for these geometrical configurations.

(i) $L = (0, a)$, $\bar{L} = (a, \infty)$

$$f_0(y) = \frac{d}{dy} \left[e^{-Ky} \int_a^y \frac{u e^{Ku}}{\left(u^2 - a^2\right)^{1/2}} du \right], \quad y \in \bar{L} = (a, \infty),$$

$$g_0(y) = e^{-Ky} \int_y^a \frac{u e^{Ku}}{\left(a^2 - u^2\right)^{1/2}} du, \quad y \in L = (0, a).$$

(ii) $L = (a, \infty)$, $\bar{L} = (0, a)$

$$f_0(y) = -\frac{d}{dy} \left[e^{-Ky} \int_y^a \frac{e^{Ku}}{\left(a^2 - u^2\right)^{1/2}} du \right], \quad y \in \bar{L} = (0, a),$$

$$g_0(y) = -e^{-Ky} \int_a^y \frac{e^{Ku}}{\left(u^2 - a^2\right)^{1/2}}, \quad y \in L = (a, \infty).$$

(iii) $L = (a,b), \quad \overline{L} = (0,a) \cup (b,\infty)$

$$f_0(y) = \begin{cases} \dfrac{d}{dy}\left[e^{-Ky} \displaystyle\int_a^y \dfrac{\left(d^2 - u^2\right)e^{Ku}}{\left|\rho(u)\right|^{1/2}}\,du \right], & 0 < y < a \\[4mm] -\dfrac{d}{dy}\left[e^{-Ky} \displaystyle\int_b^y \dfrac{\left(d^2 - u^2\right)e^{Ku}}{\left|\rho(u)\right|^{1/2}}\,du \right], & y > b \end{cases}$$

$$g_0(y) = e^{-Ky} \int_a^y \frac{\left(d^2 - u^2\right)e^{Ku}}{\left|\rho(u)\right|^{1/2}}\,du, \quad a < y < b$$

where

$$\rho(u) = \left(u^2 - a^2\right)\left(u^2 - b^2\right),$$

$$d^2 = \frac{\displaystyle\int_a^b \frac{u^2 e^{Ku}}{\left|\rho(u)\right|^{1/2}}\,du}{\displaystyle\int_a^b \frac{e^{Ku}}{\left|\rho(u)\right|^{1/2}}\,du}.$$

(iv) $L = (0,a) \cup (b,\infty), \quad \overline{L} = (a,b)$

$$f_0(y) = \frac{d}{dy}\left[e^{-Ky} \int_b^y \frac{u e^{Ku}}{\left|\rho(u)\right|^{1/2}}\,\{\delta - F(u)\}\,du \right], \quad a < y < b,$$

$$g_0(y) = \begin{cases} e^{-Ky} \displaystyle\int_a^y \dfrac{u e^{Ku}}{\left|\rho(u)\right|^{1/2}}\,\{\delta - F(u)\}\,du, & 0 < y < a, \\[4mm] e^{-Ky} \displaystyle\int_b^y \dfrac{u e^{Ku}}{\left|\rho(u)\right|^{1/2}}\,\{\delta - F(u)\}\,du, & y > b, \end{cases}$$

where

$$\delta = \frac{\displaystyle\int_a^b \frac{u e^{Ku}}{\left|\rho(u)\right|^{1/2}}\,F(u)\,du - \frac{e^{Ka}}{K}}{\displaystyle\int_a^b \frac{u e^{Ku}}{\left|\rho(u)\right|^{1/2}}\,du}$$

with

$$F(u) = \frac{2}{\pi} \int\limits_0^a \frac{|\rho(s)|^{1/2}}{u^2 - s^2} ds$$

and $\rho(u)$ is given in (iii) above.

The oblique scattering problems corresponding to the first, third and fourth configurations have been considered by Evans and Morris (1973a), Mandal and Das (1996) and Das et al (1997). For the second configuration, Evans and Morris (1973a) reported that similar single-term approximations do not provide good results. For this reason, and to obtain more accurate results, multi-term Galerkin approximations in terms of suitable basic functions are required. These are discussed in Mandal and Chakrabarti (1999) and the references given there.

7.3 GALERKIN METHOD FOR SINGULAR INTEGRAL EQUATIONS

Consider the singular integral equation

$$(I - C)\varphi = f, \tag{7.3.1}$$

written in the operator form, i.e., the equation

$$\varphi(x) - \lambda \int\limits_a^b \frac{\varphi(t)}{t - x} dt = f(x), \quad a < x < b \tag{7.3.2}$$

where

$$(I\varphi)(x) = \varphi(x), \tag{7.3.3}$$

I being the identity operator, and

$$(C\varphi)(x) = \lambda \int\limits_a^b \frac{\varphi(t)}{t - x} dt, \quad a < x < b, \tag{7.3.4}$$

C being the Cauchy singular operator, a, b being two known real numbers with $a < b$, and λ being a known complex parameter of the problem under consideration Writing

$$\varphi(x) \approx \Phi_N(x) \equiv \sum_{j=1}^N a_j \, \varphi_j(x) \tag{7.3.5}$$

where the set $\left\{\varphi_j\right\}_{j=1}^N$ is linearly independent and a_j's $(j=1,2,...,N)$ are unknown constants and using the definition of the inner product $[u,v]$ as

$$[u,v]=[v,u]^* = \int_a^b u(x)\, v^*(x)\, dx \qquad (7.3.6)$$

we obtain the following system of linear equations for the determination of the unknown constants a_j's $(j=1,2,...,N)$

$$A\alpha^{(N)} = \gamma^{(N)} \qquad (7.3.7)$$

where the elements a_{jk} of the matrix A are given by

$$a_{jk} = \delta_{jk} - \left[\lambda\varphi_j, \varphi_k\right], \; j,k = 1,2,...,N, \qquad (7.3.8)$$

and the N-dimensional vectors $\alpha^{(N)}$ and $\gamma^{(N)}$ are given by

$$\alpha^{(N)} = (a_1,a_2,...,a_N)^T \qquad (7.3.9)$$

and

$$\gamma^{(N)} = \left([f,\varphi_1],[f,\varphi_2],...,[f,\varphi_N]\right)^T \qquad (7.3.10)$$

in standard notations.

It is well known from linear algebra, because of various errors occurring in the formation of the system of equations (7.3.7) and its numerical solution, that we do not get the exact value of the solution vector $\alpha^{(N)}$ in (7.3.7), and instead, we obtain an approximate solution $\hat{\alpha}^{(N)}$ to $\alpha^{(N)}$, which is the solution of the perturbed system

$$(A+\Delta A)\hat{\alpha}^{(N)} = \gamma^{(N)} + \Delta\gamma^{(N)} \qquad (7.3.11)$$

where ΔA and $\Delta\gamma^{(N)}$ are perturbations to numerical integrations involving the various inner products.

Then, if we can associate with the perturbation ΔA, an appropriate norm, i.e. if $\|\Delta A\|$ is suitably defined, and also if $\|\Delta A\|$ is sufficiently small, it follows that

$$\left\|\alpha^{(N)} - \hat{\alpha}^{(N)}\right\| \le \|A^{-1}\| \frac{\left\|\Delta\gamma^{(N)}\right\| + \|\Delta A\|\left\|\alpha^{(N)}\right\|}{1 - \|A^{-1}\|\|\Delta A\|} \qquad (7.3.12)$$

in standard notations, where A^{-1} is the inverse of the matrix A. The inequality (7.3.12) provides us with a bound of the error in the approximation of the solution of the system of equation (7.3.7), which also gives rise to a bound of the *relative error* as given by

$$\frac{\left\|\alpha^{(N)}-\hat{\alpha}^{(N)}\right\|}{\left\|\alpha^{(N)}\right\|} \leq \frac{\left\|A^{-1}\right\|\|A\|}{1-\left\|A^{-1}\right\|\|\Delta A\|}\left[\frac{\left\|\Delta\gamma^{(N)}\right\|}{\left\|\gamma^{(N)}\right\|}+\frac{\|\Delta A\|}{\|A\|}\right]. \quad (7.3.13)$$

An excellent account of Galerkin method applied to Cauchy type singular integral equations can be found in the treatise by Golberg and Ghen (1997).

Chapter 8

Numerical Methods

In this chapter we describe briefly the development of some methods to obtain solutions of singular integral equations with Cauchy type singularities and hypersingular integral equations. Equations of both first and second kinds will be considered.

8.1 THE GENERAL NUMERICAL PROCEDURE FOR CAUCHY SINGULAR INTEGRAL EQUATION

Let us consider the singular integral equation of the form (involving a known weight function, for convenience)

$$a(x)\omega(x)\varphi(x)+b(x)\int_\alpha^\beta \frac{\omega(t)\varphi(t)}{t-x}dt - \int_\alpha^\beta k_0(x,t)\omega(t)\varphi(t)dt = f(x), \quad \alpha < x < \beta \quad (8.1.1)$$

where φ is the unknown function to be determined, and $a(x)$, $b(x)$, $\omega(x)$, $k_0(x,t)$ and $f(x)$ are all known functions for $x \in (\alpha,\beta)$ and $t \in (\alpha,\beta)$, α and β being two known real constants, to be precise. The kernel function $k_0(x,t)$ is assumed to be continuous and square integrable in (α,β) X (α,β).

If $a(x) \equiv 0$, then the equation (8.1.1) is an equation of the first kind, while if $a(x) \neq 0$, it is an equation of the second kind. Here we explain numerical methods of solution of the general singular integral equation (8.1.1), and for this purpose, we first consider the numerical approximation of integrals of the form

$$(K\varphi)(x) \equiv \int_\alpha^\beta k(x,t)\,\omega(t)\,\varphi(t)\,dt, \quad \alpha < x < \beta. \quad (8.1.2)$$

If $k(x,t)$ is continuous on $(\alpha,\beta) \times (\alpha,\beta)$, then we can approximate $(K\varphi)(x)$, for any fixed $x \in (\alpha,\beta)$, by an expression of the form

$$(K\varphi)(x) \approx \sum_{j=0}^{n} w_j \, k(x,t_j) \, \varphi(t_j) \tag{8.1.3}$$

by standard quadrature methods involving the interpolation points $\alpha = t_0 < t_1 < \dots < t_n = \beta$.

If $k(x,t)$ has singularities, then other approaches are needed and, then, if we assume that $k(x,t)$ can be factored as

$$k(x,t) = m(x,t) \, n(x,t) \tag{8.1.4}$$

where $m(x,t)$ is the singular part of $k(x,t)$ and $n(x,t)$ is continuous, then we first approximate the expression $n(x,t)\varphi(t)$ as

$$n(x,t) \, \varphi(t) \approx \sum_{i=0}^{r} a_i(t) \, \psi_i(x) \tag{8.1.5}$$

(a degenerate approximation) with some known function $\{a_i(t)\}_{i=0}^{r}$ and $\{\psi_i(x)\}_{i=0}^{r}$ and obtain

$$(K\varphi)(x) \approx \int_{\alpha}^{\beta} \omega(t) \, m(x,t) \left\{ \sum_{i=0}^{r} a_i(t) \, \psi_i(x) \right\} dt$$

$$= \sum_{i=0}^{r} w_i(x) \, \psi_i(x) \tag{8.1.6}$$

where

$$w_i(x) = \int_{\alpha}^{\beta} m(x,t) \, \omega(t) \, a_i(t) \, dt, \quad i=0,1,\dots,r. \tag{8.1.7}$$

Remarks

1. The approximation (8.1.5) is derived by interpolating $n(x,t)\varphi(t)$ in 't', in some manner—usually by a polynomial, piecewise polynomial or trigonometric interpolant.
2. If $w_i(x)$ $(i = 0,1,\dots,r)$ can be evaluated analytically or numerically to high accuracy, by some other quadrature rule, then (8.1.6) is an effective approximation formula for the integral $(K\varphi)(x)$.

As an illustration of the above idea on the numerical approximation of integrals, we take up the following example.

Example: Consider the evaluation of the weakly singular integral

$$I(x) = \int_{-1}^{1} \left(\frac{1-t}{1+t}\right)^{1/2} ln\,|x-t|\,f(t)\,dt, \quad -1 < x < 1,$$

where $f(t)$ is a smooth function. We note that $I(x)$ can be expressed as

$$I(x) = \int_{-1}^{1} \frac{(1-t)\,ln\,|x-t|\,f(t)}{\left(1-t^2\right)^{1/2}}\,dt, \quad -1 < x < 1. \qquad (8.1.8)$$

Using Chebyshev series approximation, we can write

$$(1-t)f(t) \approx \sum_{j=0}^{M}{}' c_j^{(M)}\,T_j(t) \qquad (8.1.9)$$

where $c_j^{(M)}$'s are known constants, T_j's are the Chebyshev polynomials of the first kind and the prime (') indicates that the first term is halved, we obtain

$$I(x) \approx \sum_{j=0}^{M}{}' c_j^{(M)} \int_{-1}^{1} \frac{ln\,|x-t|\,T_j(t)}{\left(1-t^2\right)^{1/2}}\,dt, \quad -1 < x < 1$$

$$(8.1.10)$$

$$= -\frac{1}{2}\,c_0^{(M)}\,ln\,2 - \sum_{j=1}^{M} \frac{c_j^{(M)}T_j(x)}{j}$$

by using the result

$$\int_{-1}^{1} \frac{ln\,|x-t|\,T_j(t)}{\left(1-t^2\right)^{1/2}}\,dt = \begin{cases} -\pi\,ln\,2, & j=0, \\ -\dfrac{\pi}{j}\,T_j(x), & j \geq 1. \end{cases}$$

For the best approximation in (8.1.9) (cf. Atkinson (1978)) $c_j^{(M)}$'s $(j = 0,1,...,M)$ are given by

$$c_j^{(M)} = \frac{2}{M+1} \sum_{i=0}^{M+1}{}'' g(\theta_i) \cos\left(\frac{ij\pi}{M+1}\right), \quad 0 \leq j \leq M \qquad (8.1.11)$$

with

$$g(t) = (1-t)f(t) \qquad (8.1.12)$$

and

$$\theta_i = \cos\frac{i\pi}{M+1}, \quad 0 \le i \le M+1 \tag{8.1.13}$$

where $\displaystyle\sum{}''$ denotes that the first and the last terms of the summation are halved.

We now take up the approximate evaluation of the singular integral

$$(S\varphi)(x) = \int_\alpha^\beta \frac{\omega(t)\varphi(t)}{t-x}\, dt, \quad \alpha < x < \beta, \tag{8.1.14}$$

occurring in the equation (8.1.1). We can easily derive the following important result, by using standard procedures in numerical analysis (cf. Atkinson (1978)).

Result: Let $\omega(t)$ be a non-negative function in the interval (α, β) such that

$$\int_\alpha^\beta t^n\,\omega(t)\,dt < \infty \quad \text{for } n \ge 0.$$

Then, using an N-point polynomial quadrature rule, we have

$$\int_\alpha^\beta \omega(t)\varphi(t)\, dt \approx \sum_{k=1}^N w_k \varphi(t_k) \tag{8.1.15}$$

where

$$w_k = \int_\alpha^\beta \omega(t)\, l_k^{(0)}(t)\, dt, \tag{8.1.16}$$

$l_k^{(0)}(t)$'s $(k = 1, 2, ..., N)$ being the fundamental polynomials of Lagrange interpolation for the set of points $t_1, t_2, ..., t_N$, i.e.

$$l_k^{(0)}(t) = \prod_{\substack{j=1 \\ j \ne k}}^N \frac{t-t_j}{t_k-t_j}.$$

We also have, under these circumstances,

$$(S\varphi)(x) \approx (S_N\varphi)(x) = \sum_{k=1}^N \frac{w_k\varphi(t_k)}{t_k-x} + \frac{q_N(x)\varphi(x)}{\sigma_N(x)}, \quad x \ne t_k,\ 1 \le k \le N, \tag{8.1.17}$$

with

$$w_k = \int_\alpha^\beta \frac{\omega(t)\,\sigma_N(t)}{\sigma'_N(t_k)(t-t_k)}\,dt,\qquad(8.1.18)$$

where $S_N\varphi = S\varphi_N$, $\varphi_N(t)$ being the unique polynomial which interpolates $\varphi(t)$ at

$$\{t_1,t_2,...,t_n,x\},\ \sigma_N(t) = \prod_{k=1}^N (t-t_k)\ \text{ and }\ q_N = S\sigma_N.$$

Proof: We first note that, by using Lagrange's interpolation formula, we have

$$\varphi_N(t) = \sum_{k=1}^N l_k(t)\varphi(t_k) + l_{N+1}(t)\varphi(x)\qquad(8.1.19)$$

where

$$l_k(t) = \frac{\sigma_N(t)(t-x)}{\sigma'_N(t_k)(t-t_k)(t_k-x)},\ 1\le k\le N$$

$$l_{N+1}(t) = \frac{\sigma_N(t)}{\sigma_N(x)}\qquad(8.1.20)$$

and the dash denoting derivative. Thus we obtain

$$\begin{aligned}(S_N\varphi)(x) &= (S\varphi_N)(x)\\[4pt]
&= \sum_{k=1}^N \varphi(t_k)\int_\alpha^\beta \frac{\omega(t)l_k(t)}{t-x}\,dt + \varphi(x)\int_\alpha^\beta \frac{\omega(t)l_{N+1}(t)}{t-x}\,dt\\[4pt]
&= \sum_{k=1}^N \varphi(t_k)\int_\alpha^\beta \frac{\omega(t)\sigma_N(t)}{\sigma'_N(t_k)(t-t_k)(t_k-x)}\,dt + \frac{(S\sigma_N)(x)}{\sigma_N(x)}\varphi(x)\\[4pt]
&= \sum_{k=1}^N \frac{w_k\varphi(t_k)}{t_k-x} + \frac{q_N(x)\varphi(x)}{\sigma_N(x)}\ \text{ for } x\ne t_k,\ 1\le k\le N,\end{aligned}\qquad(8.1.21)$$

and this proves the result (8.1.17) along with the relation (8.1.18).

Remark

The above formulae are sufficiently general, and particular cases depending on the interpolation points can be dealt with separately.

Utilizing the above results and discussions into the singular integral equation (8.1.1), we find that we can obtain the following approximate relation, connecting the $(N+1)$ unknowns $\varphi(t_k)$, $k = 1, 2, ..., N,$ where $\alpha = t_1 < t_2 < ... < t_N = \beta$ and $\varphi(x)$, as given by,

$$a(x)\omega(x)\varphi(x) + b(x)(S_N\varphi)(x) - \sum_{k=1}^{N} w_k^{(0)}(x)\varphi(t_k) = f(x), \text{ for } x \neq t_k, \; 1 \leq k \leq N \quad (8.1.22)$$

where

$$w_k^{(0)}(x) = \int_{\alpha}^{\beta} k_0(x, t_k)\, \omega(t)\, l_k^{(0)}(t)\, dt \quad (8.1.23)$$

in which $l_k^{(0)}(t)$ is given by the formula (8.1.16) and $(S_N\varphi)(x)$ is given by the formula (8.1.21).

The equation (8.1.22) can finally be expressed as, after using the equation (8.1.21)

$$b(x)\sum_{k=1}^{N} \frac{w_k \varphi(t_k)}{t_k - x} - \sum_{k=1}^{m} \omega_k^{(0)}(x)\varphi(t_k) + \left\{ a(x)\omega(x) + \frac{b(x)q_N(x)}{\sigma_N(x)} \right\}\varphi(x) = f(x). \quad (8.1.24)$$

$$\text{for } x \neq t_k, \; 1 \leq k \leq N.$$

We thus observe that, we can hope to get a system of N equations, for the N unknowns $\varphi(t_k)$ $(k = 1, 2, ..., N)$, out of the relation (8.1.24), if x can be chosen such that $x \neq t_k$ $(k = 1, 2, ..., N)$, and at the same time

$$a(x)\omega(x) + \frac{b(x)q_N(x)}{\sigma_N(x)} = 0, \quad (8.1.25)$$

and if, we can obtain compatible system of equations for the unknowns $\varphi(t_k), (1 \leq k \leq N)$, ultimately.

Thus the singular integral equation (8.1.1) can be solved, by the above explained approximate method, only under certain specific assumptions on the functions $a(x)$, $b(x)$ and $\omega(x)$. In the special case when $a(x) = 0$, we find that the values of x which can provide a solvable system of equations for the unknowns $\varphi(t_k)$ $(k = 1, 2, ..., N)$, are given by the solution of the equation

$$q_N(x) = 0 \quad (8.1.26)$$

where $q_N(x) = (S\sigma_N)(x)$, as introduced earlier. Having solved for $\varphi(t_k)(k = 1, 2, ..., N)$, the function $\varphi(x)$ for any $x \in (\alpha, \beta)$, can be constructed by using a suitable interpolation formula.

In another *special case*, when $a(x) = a$ and $b(x) = b$ where a and b are constants, along with the assumption that $\alpha = -1$, $\beta = 1$, in the given equation (8.1.1), we obtain the replacement of the result (8.1.24) by the following:

$$a\,\omega(x)\varphi(x) + b\sum_{j=0}^{n} \frac{w_j\varphi(t_j)}{t_j - x} + b\frac{(S\sigma_{n+1})(x)\varphi(x)}{\sigma_{N+1}(x)}$$

$$+ \sum_{j=0}^{n} w_j^{(0)}\varphi(t_j) = f(x), \quad \text{for } x \neq t_j,\, 0 \leq j \leq n \tag{8.1.27}$$

where $-1 = t_0 < t_1 < t_2 < ... < t_n = 1$, $\omega_j^{(0)}(x)$ being given by (8.1.23) with k replaced by j, and

$$\sigma_{n+1}(t) = \prod_{j=0}^{n} (t - t_j). \tag{8.1.28}$$

Thus, in this special case, if we can get enough number of values of x, for which the coefficient of $\varphi(x)$ in the left side of the relation (8.1.27) vanishes, i.e. if for these values of x, we have

$$a\,\omega(x)\sigma_{n+1}(x) + b(S\sigma_{n+1})(x) = 0, \tag{8.1.29}$$

we can expect to solve the resulting system of linear algebraic equations (arising out of (8.1.27))

$$b\sum_{j=0}^{n} \frac{w_j\varphi(t_j)}{t_j - x} + \sum_{j=0}^{n} w_j^{(0)}(x)\,\varphi(t_j) = f(x),\, x \neq t_j,\, 0 \leq j \leq n, \tag{8.1.30}$$

for the $(n+1)$ unknowns $\varphi(t_j)$ $(j = 0,1,...,n)$.

Note that if $b = 0$, the roots of the equation (8.1.29) are different from t_j $(j = 0,1,...,n)$.

We close this discussion here and do not go into further details of this method.

8.2 A SPECIAL NUMERICAL TECHNIQUE TO SOLVE SINGULAR INTEGRAL EQUATIONS OF FIRST KIND WITH CAUCHY KERNEL

In this section, we present a specially designed numerical scheme given by Chakrabarti and Vanden Berghe (2004) to tackle a class of singular integral equations of the first kind with Cauchy type kernels for their approximate numerical solutions. This method is guided by the analytical solutions of first kind singular integral equations with Cauchy kernel discussed in section 2.3. Let the singular integral equation of first kind with Cauchy type kernel over a finite interval, be represented by the general equation

$$\int_{-1}^{1} \{k_0(t,x) + k(t,x)\}\varphi(t)\, dt = f(x), \quad -1 < x < 1 \qquad (8.2.1)$$

where

$$k_0(t,x) = \frac{\hat{k}(t,x)}{t-x}, \qquad (8.2.2)$$

the functions $\hat{k}(t,x)$, $k(t,x)$ being regular and square integrable, $\hat{k}(x,x) \neq 0$. The integral in (8.2.1) involving the factor $(t-x)^{-1}$ is defined in the sense of Cauchy principal value. Solution of the integral equation is to be found for the following four basically important and interesting cases:

Case (i) $\varphi(x)$ is unbounded at both end points $x = \pm 1$,

Case (ii) $\varphi(x)$ is unbounded at the end point $x = -1$, but bounded at the end point $x = 1$,

Case (iii) $\varphi(x)$ is bounded at the end point $x = -1$, but unbounded at the end point $x = 1$,.

Case (iv) $\varphi(x)$ is bounded at both the end points $x = \pm 1$.

The simplest integral equation of the form (8.2.1) is the one as given by

$$\int_{-1}^{1} \frac{\varphi(t)}{t-x}\, dt = f(x), \quad -1 < x < 1 \qquad (8.2.3)$$

for which $\hat{k}(t,x) \equiv 1$ and $k(t,x) \equiv 0$. From section 2.3 it is well known that the complete analytic solutions of the singular integral equation (8.2.3), in the above four cases can be determined using the following formulae.

Case (i) $\varphi(x) = \dfrac{1}{\left(1-x^2\right)^{1/2}}\left[C - \dfrac{1}{\pi^2} \int_{-1}^{1} \dfrac{\left(1-t^2\right)^{1/2} f(t)}{t-x}\, dt \right], \quad -1 < x < 1,$ $\qquad (8.2.4)$

where C is an arbitrary constant,

Case (ii) $\varphi(x) = -\dfrac{1}{\pi^2}\left(\dfrac{1-x}{1+x}\right)^{1/2}\displaystyle\int_{-1}^{1}\left(\dfrac{1+t}{1-t}\right)^{1/2}\dfrac{f(t)}{t-x}\,dt,\quad -1<x<1,$ (8.2.5)

Case (iii) $\varphi(x) = -\dfrac{1}{\pi^2}\left(\dfrac{1+x}{1-x}\right)^{1/2}\displaystyle\int_{-1}^{1}\left(\dfrac{1-t}{1+t}\right)^{1/2}\dfrac{f(t)}{t-x}\,dt,\quad -1<x<1,$ (8.2.6)

Case (iv) $\varphi(x) = -\dfrac{\left(1-x^2\right)^{1/2}}{\pi^2}\displaystyle\int_{-1}^{1}\dfrac{f(t)}{\left(1-t^2\right)^{1/2}(t-x)}\,dt,\quad -1<x<1,$ (8.2.7)

the solution existing in *Case (iv)*, if and only if

$$\int_{-1}^{1}\frac{f(t)}{\left(1-t^2\right)^{1/2}}\,dt = 0.$$ (8.2.8)

Guided by the analytical results available, as given by the expressions (8.2.4) to (8.2.7), for the solution of the simple singular integral equation (8.2.3), as well as by utilizing the idea of replacing the integral by an appropriate approximate function, we explain here a numerical scheme that can be developed and implemented, for obtaining the approximate solutions of the general singular integral equation (8.2.1). The particular cases of the equation (8.2.3) follow quite easily and the known analytical solutions are recovered in the cases of simple forms of the following function $f(x)$, being polynomials of low degree.

The numerical scheme:

The unknown function $\varphi(x)$ is represented as

$$\varphi(x) = \frac{\psi(x)}{\left(1-x^2\right)^{1/2}},\quad -1<x<1,$$ (8.2.9)

where $\psi(x)$ is a well behaved function of $x \in [-1,1]$. Then the unknown function $\psi(x)$ is approximated by means of a polynomial of degree n, as given by

$$\psi(x) \equiv \psi_r(x) \approx \sum_{j=0}^{n}\left(c_j^{(r)}x^j\right)\lambda_r(x),\quad r=1,2,3,4$$ (8.2.10)

in the four cases mentioned above, where $\lambda_1(x)=1$ in *case (i)*, $\lambda_2(x)=1-x$ in *case (ii)*, $\lambda_3(x)=1+x$ in *case (iii)* and $\lambda_4(x)=1-x^2$ in *case (iv)*; where the unknown coefficients $c_j^{(r)}$ can be determined in terms of the values of the unknown function $\psi(x)$ at $n+1$ distinct points x_j, such that $-1\le x_0<x_1<...<x_n\le 1$.

Using the above approximate form (8.2.10) of the function $\psi(x)$ along with the representation (8.2.9), into the original integral equation (8.2.1), we obtain

$$\sum_{j=0}^{n} c_j^{(r)}\left[\int_{-1}^{1} \frac{\lambda_r(t)\hat{k}(t,x)t^j}{\left(1-t^2\right)^{1/2}(t-x)}dt + \int_{-1}^{1} \frac{\lambda_r(t)k(t,x)t^j}{\left(1-t^2\right)^{1/2}}dt\right]=f(x),\ -1<x<1,\ r=1,2,3,4. \quad (8.2.11)$$

In the equations (8.2.11) we next use the following polynomial approximations to the kernels $\hat{k}(t,x)$ and $k(t,x)$ given by

$$\hat{k}(t,x)=\sum_{p=0}^{m} \hat{k}_p(x)t^p,\quad k(t,x)=\sum_{q=0}^{s} k_q(x)t^q \qquad (8.2.12)$$

with

$$\hat{k}_p(x)=\hat{k}(t_p,x),\quad -1\le t_0<t_1<...<t_m\le 1,$$
$$\hat{k}_q(x)=k(t_q',x),\quad -1\le t_0'<t_1'<...<t_s'\le 1. \qquad (8.2.13)$$

We thus obtain the following functional relation to be solved for the unknown constants $c_j^{(r)}$ $(j=0,1,...n)$

$$\sum_{j=0}^{n} c_j^{(r)}\left[\sum_{p=0}^{m} \hat{k}_p(x)\int_{-1}^{1} \frac{t^{p+j}\lambda_r(t)}{\left(1-t^2\right)^{1/2}(t-x)}dt+\sum_{q=0}^{s} k_q(x)\int_{-1}^{1} \frac{t^{q+j}\lambda_r(t)}{\left(1-t^2\right)^{1/2}}dt\right] \qquad (8.2.14)$$
$$=f(x),\ -1<x<1.$$

Now using the following well known results given by

$$\int_{-1}^{1} \frac{t^{p+j}\lambda_r(t)}{\left(1-t^2\right)^{1/2}(t-x)}dt=\pi\ x^{p+j-1}\lambda_r(x)\left(1-\frac{1}{x^2}\right)^{-1/2} \qquad (8.2.15)$$
$$\equiv u_{p+j}^{(r)}(x),\ -1<x<1$$

and

$$\int_{-1}^{1} \frac{t^{q+j} \lambda_r(t)}{\left(1-t^2\right)^{1/2}} \, dt \equiv \gamma_{q+j}^{(r)} \qquad (8.2.16)$$

where $\gamma_{q+j}^{(s)}$ is a constant obtainable in terms of Beta functions, we obtain from (8.2.14)

$$\sum_{j=0}^{n} c_j^{(r)} \left[\sum_{p=0}^{m} \hat{k}_p(x) \, u_{p+j}^{(r)}(x) + \sum_{q=0}^{s} k_q(x) \, \gamma_{q+j}^{(r)} \, x \right] = f(x), \quad -1 < x < 1, \ r = 1,2,3,4. \quad (8.2.17)$$

Setting $x = x_l$, $l = 0,1,2,...,n$ in the above relation (8.2.17), we obtain the following $(n+1) \times (n+1)$ linear equations for the determination of the unknown constants $c_j^{(r)}$ $(j = 0,1,...,n)$:

$$\sum_{j=0}^{n} c_j^{(r)} \, \alpha_{jl}^{(r)} = f_l, \quad l = 0,1,...,n; \ r = 1,2,3,4, \qquad (8.2.18)$$

where

$$f_l = f(x_l) \qquad (8.2.19)$$

and

$$\alpha_{jl}^{(r)} = \sum_{p=0}^{m} \hat{k}_p(x_l) \, u_{p+j}^{(r)}(x_l) + \sum_{q=0}^{s} k_q(x_l) \, \gamma_{q+j}^{(r)}. \qquad (8.2.20)$$

Solving the system of equations (8.2.18) and utilizing the relations (8.2.9) and (8.2.10), we determine an approximate solution of the singular integral equation (8.2.1), in the form

$$\varphi(x) \approx \lambda_r(x) \sum_{j=0}^{n} \frac{c_j^{(r)} x^j}{\left(1-x^2\right)^{1/2}}, \quad r = 1,2,3,4. \qquad (8.2.21)$$

Particular cases and examples

Example 1

For the simplest integral equation (8.2.3), we select in the *first* instance the known function $f(x)$ to be a polynomial of degree one, i.e.

$$f(x) = b_0 + b_1 x, \quad -1 < x < 1 \qquad (8.2.22)$$

with b_0 and b_1 as known constants. We first observe that due to (8.2.8) we must have for the *case (iv)* that

$$b_0 = 0. \tag{8.2.23}$$

whereas b_0 can be a non-zero constant for the other three *cases (i), (ii)* and *(iii)*, for the existence of solutions. Then, using the facts that, for the integral equation (8.2.3),

$$\hat{k}(t,x) = 1 \quad \text{and} \quad k(t,x) = 0 \tag{8.2.24}$$

along with the relations (8.2.12) to (8.2.16), we obtain that

$$\alpha_{jl}^{(r)} = u_j^{(r)}(x_l) = \pi \, x_l^{j-1} \lambda_r(x_l) \left(1 - \frac{1}{x_l^2}\right)^{-1}, \quad r = 1,2,3,4; \; l = 0,1,...,n, \tag{8.2.25}$$

and the equations (8.2.17) reduce to the simple polynomial relations

$$\sum_{j=0}^{n} c_j^{(r)} u_j^{(r)}(x) = b_0 + b_1 x, \quad r = 1,2,3,4, \tag{8.2.26}$$

so that we can determine the unknown constants $c_j^{(r)}$ directly, by just comparing the coefficients of various powers of x from both sides of the equation (8.2.26). There is thus no need to solve the system of equations (8.2.18) in this simple situation. Following expressions are easily found:

$$u_0^{(1)} = 0, \quad u_1^{(1)} = \pi, \quad u_2^{(1)} = \pi x, \quad u_3^{(1)} = \pi\left(\frac{1}{2} + x^2\right),...;$$

$$u_0^{(2)} = -\pi, \quad u_1^{(2)} = \pi(1-x), \quad u_2^{(2)} = \pi\left(-\frac{1}{2} + x - x^2\right), \quad u_3^{(2)} = \pi\left(\frac{1}{2} - \frac{x}{2} + x^2 - x^3\right),...; \tag{8.2.27}$$

$$u_0^{(3)} = \pi, \quad u_1^{(3)} = \pi(1+x), \quad u_2^{(3)} = \pi\left(\frac{1}{2} + x + x^2\right), \quad u_3^{(3)} = \pi\left(\frac{1}{2} + \frac{x}{2} + x^2 + x^3\right),...;$$

The constants $c_j^{(r)}$ can then be determined easily and the final forms of the unknown function $\varphi(x)$ agree with the known results obtainable from the relations (8.2.4) to (8.2.7).

Let us illustrate this for $r = 4$. We find from equation (8.2.26)

$$c_0^{(4)} u_0^{(4)}(x) + c_1^{(4)} u_1^{(4)}(x) + c_2^{(4)} u_2^{(4)}(x) + \cdots = b_0 + b_1 x$$

which is equivalent to

$$c_0^{(4)}(-\pi x) + c_1^{(4)}\left\{-\pi\left(x^2 - \frac{1}{2}\right)\right\} + c_2^{(4)}\left\{-\pi\left(x^3 - \frac{x}{2}\right)\right\} + ... = b_0 + b_1 x.$$

By equating similar powers of x from both sides of the equation and taking account of (8.2.18), we obtain

$$\frac{\pi}{2} c_1^{(4)} = b_0 \equiv 0.$$

$$-\pi c_0^{(4)} + \frac{\pi}{2} c_2^{(4)} = b_1,$$

$$c_j^{(4)} = 0, \quad j = 2,3,...$$

giving as a result

$$c_0^{(4)} = -\frac{b_1}{\pi}, \quad c_j^{(4)} = 0, \quad j = 1,2... .$$

The result for $\varphi(x)$ follows from (8.2.9), (8.2.10) and $\lambda_4(x) = 1 - x^2$, i.e.

$$\varphi(x) = -\frac{b_1}{\pi} \left(1 - x^2\right)^{1/2},$$

which is also the exact value obtained by (8.2.7) for $f(x)$ given by (8.2.22).

Example 2

As a *second* example we consider the integral equation

$$\int_{-1}^{1} \left(\frac{1}{t-x} + t + x\right)\varphi(t)\, dt = f(t), \quad -1 < x < 1. \qquad (8.2.28)$$

For this, we have

$$\hat{k}(t,x) = 1 \text{ and } k(t,x) = t + x.$$

So, we obtain, in this case

$$\hat{k}_0(x) = 1, \quad \hat{k}_j(x) = 0, \quad j = 1,2,...;$$

$$k_0(x) = x, \quad k_1(x) = 1, \quad k_j(x) = 0, \quad j = 2,3,...;$$

$$u_j^{(r)}(x) = \int_{-1}^{1} \frac{t^j \lambda_r(t)}{\left(1-t^2\right)^{1/2}(t-x)}\, dt, \quad j=0,1,2,...;$$

$$\gamma_j^{(r)} = \int_{-1}^{1} \frac{t^j \lambda_r(t)}{\left(1-t^2\right)^{1/2}}\, dt, \quad j=0,1,2,... .$$

Let us consider in detail the case $r = 1$. Then

$$\gamma_j^{(1)} = \int_{-1}^{1} \frac{t^j}{\left(1-t^2\right)^{1/2}} \, dt, \quad j=0,1,2,...$$

so that

$$\gamma_0^{(1)} = \pi, \quad \gamma_1^{(1)} = 0, \quad \gamma_2^{(1)} = \frac{\pi}{2}, \quad \gamma_3^{(1)} = 0, \quad \gamma_4^{(1)} = \frac{3}{8}\pi .... \quad (8.2.29)$$

Again,

$$u_j^{(1)}(x) = \int_{-1}^{1} \frac{t^j}{\left(1-t^2\right)^{1/2}(t-x)} \, dt, \quad -1 < x < 1$$

so that

$$u_0^{(1)}(x) = 0, \quad u_1^{(1)}(x) = \pi, \quad u_2^{(1)}(x) = \pi x, \quad u_3^{(1)}(x) = \pi\left(\frac{1}{2}+x^2\right),.... \quad (8.2.30)$$

Then, using (8.2.20) we get for $r = 1$,

$$\alpha_{jl}^{(1)} = u_j^{(1)}(x_l) + x_l \gamma_j^{(1)} + \gamma_{j+1}^{(1)}, \quad j,l = 0,1,2,... .$$

For $j = 0,1,,2,3,...$ we find

$$\alpha_{0l}^{(1)} = \gamma_0^{(1)} x_l + \gamma_1^{(1)},$$

$$\alpha_{1l}^{(1)} = \pi + \gamma_1^{(1)} x_l + \gamma_2^{(1)},$$

$$\alpha_{2l}^{(1)} = \pi x_l + \gamma_2^{(1)} x_l + \gamma_3^{(1)}, \quad (8.2.31)$$

$$\alpha_{3l}^{(1)} = \pi\left(\frac{1}{2}+x_l^2\right) + \gamma_3^{(1)} x_l + \gamma_4^{(1)},$$

$$. . . . .$$

By introducing (8.2.29) we obtain

$$\alpha_{0l}^{(1)}(x) = \pi x_l, \quad \alpha_{1l}^{(1)} = \frac{3}{2}\pi, \quad \alpha_{2l}^{(1)}(x) = \frac{3}{2}\pi x_l, \quad \alpha_{3l}^{(1)}(x) = \pi\left(\frac{3}{8}+x_l\right),.... \quad (8.2.32)$$

Finally, by choosing $n = 3$, we have to solve the linear system (8.2.18) for *case (i)* ($r = 1$), i.e.

$$\sum_{j=0}^{3} c_j^{(1)} \alpha_{jl}^{(1)} = f_l, \quad l = 0,1,2,3. \quad (8.2.33)$$

In the special situation, when $f(x) = 1$, the integral equation (8.2.28) can be expressed as

$$\int_{-1}^{1} \frac{\varphi(t)}{t-x} \, dt = (1 + \mu_0) + \mu_1 x, \quad -1 < x < 1, \qquad (8.2.34)$$

with

$$\mu_0 = -\int_{-1}^{1} t \, \varphi(t) \, dt, \quad \mu_1 = -\int_{-1}^{1} \varphi(t) \, dt. \qquad (8.2.35)$$

We can easily solve the equation (8.2.34), by utilizing the relation (8.2.4), in the *case (i)*, and we find that $\varphi(x)$ is given by

$$\varphi(x) = \frac{B_0}{\left(1 - x^2\right)^{1/2}} + \frac{2}{3\pi}(x - B_0 \pi x^2) \qquad (8.2.36)$$

where B_0 is an arbitrary constant.

Again, by utilizing the system of equations (8.2.33), with $g_l = 1$, for $l = 0,1,2,3$ we obtain that, for any four chosen values of $x_l (-1 \le x_l \le 1)$,

$$c_0^{(1)} = B_0, \quad c_1^{(1)} = \frac{2}{3\pi}, \quad c_2^{(1)} = -\frac{2}{3} B_0, \quad c_3^{(1)} = 0 \qquad (8.3.37)$$

where B_0 is an arbitrary constant.

Finally, by using the constants in (8.2.37) in the relation (8.3.10), in the *case (i)*, i.e. for $r = 1$, we obtain the same expression (8.2.36), which is the exact solution for the integral equation (8.2.28), in the special situation when $f(x) = 1$.

Remark

1. The reason behind the matching of our approximate solution with the exact ones, of the two singular integral equations considered above, is that the known function $f(x)$ is chosen to be a polynomial and that we take here sufficiently large n in the approximation (8.2.10).

2. We have not taken up any other example which require the solution of the system of equations (8.2.18) seriously, but the two examples taken up here assure the correctness and robustness of the method described above.

8.3 NUMERICAL SOLUTION OF HYPERSINGULAR INTEGRAL EQUATIONS USING SIMPLE POLYNOMIAL EXPANSION

A general hypersingular integral equation of first kind, over the interval $[-1,1]$, can be represented by

$$\int_{-1}^{1} \psi(t)\left\{\frac{K(t,x)}{(t-x)^2} + L(t,x)\right\} dt = f(x), \quad -1 \leq x \leq 1 \quad (8.3.1)$$

with $\varphi(\pm 1) = 0$, where $K(t,x)$ and $L(t,x)$ are regular square-integrable functions of t and x, and $K(x,x) \neq 0$, and the integral involving $(t-x)^{-2}$ is understood in the sense of Hadamard finite part.

A somewhat less general form of first kind hypersingular integral equation given by

$$\int_{-1}^{1} \varphi(t)\left\{\frac{1}{(t-x)^2} + L(t,x)\right\} dt = f(x), \quad -1 \leq x \leq 1 \quad (8.3.2)$$

with $\varphi(\pm 1) = 0$, arises in a variety of mixed boundary value problems in mathematical physics such as water wave scattering or radiation problems involving thin submerged plates (cf. Parsons and Martin (1992, 1994), Martin et al. (1997), Mandal et al. (1995), Banerjea et al. (1996), Mandal and Gayen (2002), Kanoria and Mandal(2002), Maiti and Mandal (2010)), and fracture mechanics (Chan et al. (2003)), etc. The integral equation (8.3.2) is usually solved approximately by an expansion-collocation method, the expansion being in terms of a finite series involving Chebyshev polynomials $U_i(t)$ of the second kind. In particular, $\varphi(t)$ in (8.3.2) is approximated as

$$\varphi(t) \approx \left(1-t^2\right)^{1/2} \sum_{i=0}^{n} a_i U_i(t) \quad (8.3.3)$$

where a_i $(i = 0,1,...,n)$ are unknown constants. Substitution of (8.3.3) in (8.3.1) produces

$$\sum_{i=0}^{n} a_i A_i(x) = f(x), \quad -1 \leq x \leq 1 \quad (8.3.4)$$

where

$$A_i(x) = -\pi(i+1)U_i(x) + \int_{-1}^{1} \left(1-t^2\right)^{1/2} L(t,x) U_i(t) \, dt. \quad (8.3.5)$$

To find the unknown constants a_i $(i = 0,1,...,n)$, we put $x = x_j$ $(j = 0,1,...,n)$ where x_j's are suitable collocation points such that $-1 \leq x_j \leq 1$. This produces the linear system

$$\sum_{i=0}^{n} a_i A_{ij} = f_j, \quad j=0,1,...,n \qquad (8.3.6)$$

where $A_{ij} = A_i(x_j)$ and $f_j = f(x_j)$. This can be solved by any standard method. The collocation points are usually chosen to be the zeros of $U_{n+1}(x)$ or $T_{n+1}(x)$. This method becomes somewhat unsuitable for solving the general hypersingular integral equation (8.3.1) due to presence of the factor $K(t,x)$ with $(t-x)^{-2}$.

Mandal and Bera (2006) developed a modified method to solve approximately the equation (8.4.1). This method is somewhat similar to the method described in section 8.2 to solve a general type of first kind singular integral equation with Cauchy type kernel, given by

$$\int_{-1}^{1} \varphi(t) \left\{ \frac{K(t,x)}{t-x} + L(t,x) \right\} dt = f(x), \quad -1 < x < 1, \quad (8.3.7)$$

$\varphi(t)$ satisfying appropriate conditions at the end points, and the integral involving $(t-x)^{-1}$ is in the sense of Cauchy principal value $(K(x,x) \neq 0)$. The approximate method developed below appears to be quite appropriate to solve the most general type of first kind hypersingular integral equation (8.3.1) assuming of course that $K(t,x)$ and $L(t,x)$ can be approximated as in section 8.2.

Method of solution

The unknown function $\varphi(x)$ satisfying $\varphi(\pm 1) = 0$ can be represented in the form

$$\varphi(x) = \left(1-x^2\right)^{1/2} \psi(x), \quad -1 \leq x \leq 1 \qquad (8.3.8)$$

where $\psi(x)$ is a well-behaved unknown function of $x \in [-1,1]$. Approximating $\psi(x)$ by means of a polynomial of degree n, given by

$$\psi(x) \approx \sum_{j=0}^{n} c_j x^j \qquad (8.3.9)$$

where c_j's $(j = 0,1,...,n)$ are unknown constants, then the original integral equation (8.3.1) produces

$$\sum_{j=0}^{n} c_j \left[\frac{\left(1-t^2\right)^{1/2} K(t,x)t^j}{(t-x)^2} dt + \int_{-1}^{1} \left(1-t^2\right)^{1/2} L(t,x)t^j \ dt \right] = f(x), \quad -1 \le x \le 1. \quad (8.3.10)$$

As in section 8.2, the functions $K(t,x)$ and $L(t,x)$ can be approximated as (for fixed x)

$$K(t,x) = \sum_{p=0}^{m} K_p(x)t^p, \quad L(t,x) = \sum_{q=0}^{s} L_q(x) \, t^q \quad (8.3.11)$$

with known expressions for $K_p(x)$ and $L_q(x)$. Then (8.3.10) gives

$$\sum_{j=0}^{n} c_j \alpha_j(x) = f(x), \quad -1 \le x \le 1 \quad (8.3.12)$$

where

$$\alpha_j(x) = \sum_{p=0}^{m} u_{p+j}(x) \, K_p(x) + \sum_{q=0}^{s} \gamma_{q+j}(x) \, L_q(x) \quad (8.3.13)$$

with

$$u_{p+j}(x) = \int_{-1}^{1} \frac{\left(1-t^2\right)^{1/2} t^{p+j}}{(t-x)^2} dt, \quad -1 \le x \le 1 \quad (8.3.14)$$

$$\gamma_{q+j}(x) = \int_{-1}^{1} \left(1-t^2\right)^{1/2} t^{q+j} dt, \quad (8.3.15)$$

which can be easily evaluated. The unknown constants c_j $(j = 0,1,...,n)$ are now obtained by putting $x = x_l$ $(l = 0,1,...,n)$ in (8.3.12), where $-1 \le x_l \le 1$ and are to be chosen suitably. Thus we obtain a system of $(n+1)$ linear equations, given by

$$\sum_{j=0}^{n} c_j \alpha_{jl} = f_l, \quad l = 0,1,...,n \quad (8.3.16)$$

where

$$\alpha_{jl} = \alpha(x_l), \quad f_l = f(x_l), \quad (8.3.17)$$

for the determination of the $(n+1)$ unknowns c_j $(j = 0,1,...,n)$. This completes the description of the approximate method for solving (8.3.1). Below we give some simple examples to illustrate the method.

Example 1: If we choose $K(t,x) \equiv 1$, $L(t,x) \equiv 0$, then the equation (8.3.1) reduces to the simple hypersingular integral equation

$$\int_{-1}^{1} \frac{\varphi(t)}{(t-x)^2} dt = f(x), \quad -1 < x < 1 \qquad (8.3.18)$$

satisfying $\varphi(\pm 1) = 0$, whose solution is given by (see section (5.3))

$$\varphi(t) = \frac{1}{\pi^2} \int_{-1}^{1} f(x) \ln \left| \frac{x-t}{1-xt + \left\{(1-x^2)(1-t^2)\right\}^{1/2}} \right| dx, \quad -1 \le t \le 1. \qquad (8.3.19)$$

However, here we use the method developed above to obtain the solution for the particular forcing function $f(x) = 1$. Since for this case, $K_p(x)$ and $L_p(x)$ in (8.3.11) are given by

$$K_0(x) = 1, \quad K_p(x) = 0 \ (p > 0) \quad \text{and} \quad L_q(x) = 0 \ (q \ge 0) \ (8.3.20)$$

we find that the relation (8.3.12) produces

$$\sum_{j=0}^{n} c_j u_j(x) = 1, \quad -1 \le x \le 1 \qquad (8.3.21)$$

where

$$u_0(x) = -\pi, \quad u_1(x) = -2\pi x, \quad u_2(x) = \pi \left(\frac{1}{2} - 3x^2 \right),$$

$$u_3(x) = \pi \left(x - 4x^3 \right), \quad u_4(x) = \pi \left(\frac{1}{8} + \frac{3}{2} x^2 - 5x^4 \right), \dots . \qquad (8.3.22)$$

Substituting (8.3.22) in (8.3.21) and comparing the coefficients in both sides, we obtain

$$c_0 = -\frac{1}{\pi}, \quad c_1 = c_2 = \dots = 0 \qquad (8.3.23)$$

so that

$$\varphi(x) = -\frac{1}{\pi} \left(1 - x^2 \right)^{1/2}$$

which is in fact the exact solution of (8.3.18) for $f(x) = 1$ obtained by using the relation (8.3.19).

Example 2: Next we consider the hypersingular integral equation

$$\int_{-1}^{1} \left\{ \frac{1}{(t-x)^2} + (t+x) \right\} \varphi(t)\, dt = f(x), \quad -1 \le x \le 1 \quad (8.3.24)$$

with $\varphi(\pm 1) = 0$. This corresponds to $K(t,x) \equiv 1$ and $L(t,x) \equiv t + x$.

The equation (8.3.24) is of the form (8.3.1), and as such, we use the method developed above to obtain approximate solution of (8.3.24). Now, here, $K_p(x)$ and $L_p(x)$ are given by

$$K_0(x) = 1, \quad K_p(x) = 0 \ (p \ge 1) \text{ and } L_0(x) = x, \quad L_1(x) = 1, \quad L_q(x) = 0 \ (q \ge 2). \quad (8.3.25)$$

Thus (8.3.13) gives

$$\alpha_j(x) = u_j(x) + \gamma_j x + \gamma_{j+1}, \quad j = 0,1,..., \quad (8.3.26)$$

where $u_j(x)$ $(j = 0,1,...)$ are the same as those given in (8.3.22) and

$$\gamma_{2j+1} = 0, \quad \gamma_{2j} = \frac{\pi^{1/2}\, \Gamma\left(j + \dfrac{1}{2}\right)}{2(j+1)!}, \quad j = 0,1,... \quad (8.3.27)$$

so that $\alpha_0(x), \alpha_1(x),...$ etc. are obtained in closed forms.

For simplicity if we choose the forcing function $f(x)$ to be of the form $f(x) = b_0 + b_1 x$, where b_0 and b_1 are known constants, then we can determine the unknown constants $c_0, c_1,...$ directly by comparing the coefficients of various powers of x in the two sides of (8.3.12), as both sides are now polynomials. This produces

$$c_0 = -\frac{2}{3\, l\pi}(16 b_0 + b_1), \quad c_1 = -\frac{16}{3\, l\pi}\left(\frac{b_0}{2} + b_1\right), \quad c_j = 0 \ (j \ge 2) \quad (8.3.28)$$

and the solution of (8.3.24) in this case is obtained as

$$\varphi(x) = (c_0 + c_1 x)\left(1 - x^2\right)^{1/2}. \quad (8.3.29)$$

However, if we use the expansion of $\varphi(x)$ in terms of Chebyshev polynomials given by (8.3.3), then in this case the functions $A_i(x)$ $(i = 0,1,...)$ are obtained as

$$A_0(x) = -\pi + \frac{\pi}{2}x, \ A_1(x) = \frac{\pi}{4} - 4\pi x, \ A_2(x) = 3\pi - 16\pi x^2, \ A_3(x) = 16\pi x^2 - 32\pi x^3... \quad (8.3.30)$$

and comparing the coefficients of both sides of the relation (8.3.4) we obtain

$$a_0 = -\frac{2}{3\,l\pi}(16b_0 + b_1), \ a_1 = -\frac{8}{3\,l\pi}\left(\frac{b_0}{2} + b_1\right), \ a_i = 0 \ (i \geq 2). \quad (8.3.31)$$

Noting that $U_0(x) = 1$ and $U_1(x) = 2x$, we find from (8.3.3) that $\varphi(x)$ is exactly the same as given in (8.3.29).

It may be noted that the collocation method to obtain the unknown constants c_i $(i = 0,1,...)$ in (8.3.9) and a_i $(i = 0,1,...)$ in (8.3.3) for this problem can be used. For simplicity we choose $f(x) = 1 + 2x$ so that $b_0 = 1$ and also $b_1 = 2$ above. Choosing $n = 3$ in the expansion (8.3.4). The unknown constants c_0, c_1, c_2, c_3 are determined from the linear system

$$\sum_{j=0}^{3} c_j \alpha_{jl} = f_l, \ l = 0,1,2,3. \quad (8.3.32)$$

If we choose the collocation points as $x_0 = -1$, $x_1 = -\frac{1}{3}$, $x_2 = \frac{1}{3}$, $x_3 = 1$, then the linear equations (8.3.32) produce

$$c_0 = -0.3696501, \ c_1 = -0.4107224, \ c_2 \approx 0, \ c_3 \approx 0, \ (8.3.33)$$

which are almost the same as given in (8.3.28). Similarly choosing $n = 3$ in (8.3.3), we see that the unknown constants a_0, a_1, a_2, a_3 are to be found by solving the linear system

$$\sum_{i=0}^{3} a_i A_{ij} = f_j, \ j = 0,1,2,3 . \quad (8.3.34)$$

Choosing the same set of collocation points as -1, $-\frac{1}{3}$, $\frac{1}{3}$, 1, we find that the linear equations (8.3.34), when solved, produce

$$a_0 = -0.3696500, \ a_1 = -0.2053610, \ a_2 \approx 0, \ a_3 \approx 0 \ (8.3.35)$$

which are again almost the same as given in (8.3.31). It may be noted that by increasing n, the same results as above are obtained for both the methods.

8.4 NUMERICAL SOLUTION OF SIMPLE HYPERSINGULAR INTEGRAL EQUATION USING BERNSTEIN POLYNOMIALS AS BASIS

The simple hypersingular integral equation (8.3.18) has also been solved numerically by Mandal and Bhattacharya (2007) employing the Bernstein polynomials as basis for the expansion of the unknown function $\varphi(x)$. This is briefly presented here. We can write

$$\varphi(x) = \left(1 - x^2\right)^{1/2} \psi(x), \quad -1 \le x \le 1 \tag{8.4.1}$$

where $\psi(x)$ is a well-behaved function of $x \in [-1,1]$. Now $\psi(x)$ is approximated in terms of Bernstein polynomials in the form

$$\psi(x) = \sum_{l=0}^{n} a_i B_{i,n}(x), \quad -1 \le x \le 1. \tag{8.4.2}$$

Then the equation (8.3.16) produces the relation

$$\sum_{i=0}^{n} a_i A_i(x) = f(x), \quad -1 \le x \le 1 \tag{8.4.3}$$

where

$$A_i(x) = \frac{1}{2^n} \binom{n}{i} \sum_{k=0}^{n} d_k^{i,n} \left[-\pi(k+1)x^k + \sum_{m=0}^{k-2} \frac{1+(-1)^m}{4} \frac{\Gamma\left(\frac{m+1}{2}\right)\Gamma\left(\frac{1}{2}\right)}{\Gamma\left(\frac{m}{2}+2\right)} (k-m-1)x^{k-m-2} \right], \tag{8.4.4}$$

the summation inside the square bracket being understood to be absent for $k < 2$, and $d_k^{i,n}$'s are given in (6.5.16) of section 6.5.

The unknown constants a_i $(i = 0,1,...,n)$ can be found by a collocation method as has been done in section 8.3 where an expansion for $\psi(x)$ has been used in terms of simple polynomials instead of Bernstein polynomials. However, here we follow a different method as described below.

Multiplying both sides of (8.4.3) by $B_{j,n}(x)$ $(j = 0,1,...,n)$ and integrating with respect to x between -1 to 1, we obtain

$$\sum_{i=0}^{n} a_i c_{ij} = f_j, \quad j = 0,1,...n \tag{8.4.5}$$

where now

$$c_{ij} = \int_{-1}^{1} A_i(x)\, B_{j,n}(x)\, dx$$

$$= \frac{1}{2^{2n}} \binom{n}{i}\binom{n}{j} \sum_{k=0}^{n} \sum_{r=0}^{n} d_k^{i,n}\, d_r^{j,n} \left[-\pi(k+1)\frac{1+(-1)^{k+r}}{k+r+1} \right. \tag{8.4.6}$$

$$\left. + \sum_{m=0}^{k-2} \frac{1+(-1)^{k+r-m}}{k+r-m-1}\, \frac{1+(-1)^m}{4}\, \frac{\Gamma\!\left(\frac{1}{2}\right)\Gamma\!\left(\frac{m+1}{2}\right)}{\Gamma\!\left(\frac{m}{2}+2\right)}(k-m-1) \right]$$

and

$$f_j = \int_{-1}^{1} f(x)\, B_{j,n}(x)\, dx, \quad j = 0,1,...,n. \tag{8.4.7}$$

We note that when $f(x) = 1$,

$$f_j = \frac{1}{2^n}\, \binom{n}{j} \sum_{k=0}^{n} \frac{1+(-1)^k}{1+k}\, d_k^{j,n}, \quad j = 0,1,...,n. \tag{8.4.8}$$

The constants $d_k^{i,n}$ occurring in (8.4.4) and (8.4.8) are defined in (6.5.16) of section 6.5.

In our numerical computation here, $f(x)$ is chosen to be 1, and n to be 3. The constants a_i $(i = 0,1,2,3)$ are calculated by solving the linear system (8.4.5) for $n = 3$ and f_j $(j = 0,1, 2, 3)$ given by (8.4.8). Thus the function $\psi(x)$ is found approximately, and hence, by using the relation (8.4.1), $\varphi(x)$ is obtained approximately. A comparison between this

approximate solution and the exact solution given by

$$\varphi(x) = -\frac{1}{\pi}(1-x^2)^{1/2}$$

is presented in Table 1 for $x = 0$, ± 0.2, ± 0.4, ± 0.6, ± 0.8. It is seen that the approximate and the exact values are same and they coincide. Mandal and Bhattacharya (2007) plotted the absolute difference between exact and approximate solutions and found from that figure that the accuracy is of the order of 10^{-17}.

Table 1

x	0	± 0.2	± 0.4	± 0.6	± 0.8
$\varphi(x)$ (approx)	-0.318310	-0.311879	-0.291736	- 0.254648	-0.190986
$\varphi(x)$ (exact)	-0.318310	-0.311879	-0.291736	- 0.254648	-0.190986

Convergence of the method

The simple hypersingular integral equation (8.3.18) has the representation in the operator form

$$(H\psi)(x) = f_1(x), \quad -1 \leq x \leq 1 \tag{8.4.9}$$

where H is the operator defined by

$$(H\psi)(x) = \frac{1}{\pi} \frac{d}{dx}\left[\int_{-1}^{1} \frac{\left(1-t^2\right)^{1/2}}{t-x}\psi(t)\, dt \right], \quad -1 \leq x \leq 1, \tag{8.4.10}$$

the integral within the square bracket being in the sense of Cauchy principal value, and

$$f_1(x) = \frac{1}{\pi} f(x). \tag{8.4.11}$$

Since

$$(HU_n)(x) = -(n+1)\, U_n(x), \quad n \geq 0,$$

H can be extended as a bounded linear operator (cf. Golberg and Chen (1007, p 306)) from $L_1(w)$ to $L(w)$, where $L(w)$ is the space of functions square integrable with respect to the weight $w(x) = \left(1-x^2\right)^{1/2}$ in $[-1,1]$, and $L_1(w)$ is the subspace of functions $u \in L(w)$ satisfying

$$\|u\|_1^2 = \sum_{k=0}^{\infty} (k+1) < u,\, u_k >_w^2 < \infty \tag{8.4.12}$$

where

$$< u, u_k >_w = \int_{-1}^{1} \left(1-x^2\right)^{1/2} u(x)\, u_k(x)\, dx. \tag{8.4.13}$$

Now the function $\psi(x)$ satisfying the equation (8.4.9) is approximated in terms of the Bernstein polynomials $B_{i,n}(x)$ in the form

$$\psi(x) \simeq p_n(x) \qquad (8.4.14)$$

where

$$p_n(x) = \sum_{i=0}^{n} a_i \, B_{i,n}(x) \qquad (8.4.15)$$

In terms of the orthonormal Chebyshev polynomials $u_j(x) = \sqrt{\dfrac{2}{\pi}} \, U_j(x)$, $p_n(x)$ can be expressed in the form

$$p_n(x) = \sum_{j=0}^{n} c_j u_j(x) \qquad (8.4.16)$$

where $c_j = \sqrt{\dfrac{\pi}{2}} \, b_j$, b_j $(j = 0,1,...,n)$ being expressed in terms of a_i $(i = 0,1,...,n)$. If $f_1 \in C^r[-1,1]$, $r > 0$, then it follows that (cf. Golberg and Chen (1997, p 306)).

$$\left\| \psi - p_n \right\|_1 \leq c_1 \, n^{-r} \qquad (8.4.17)$$

where c_1 is a constant. Thus the convergence is quite fast if r is large. Here we have chosen f_1 to be a constant and thus $f_1 \in C^\infty[-1,1]$. Hence the convergence is very rapid and this has been reflected in the numerical computation.

8.5 NUMERICAL SOLUTION OF SOME CLASSES OF LOGARITHMICALLY SINGULAR INTEGRAL EQUATIONS USING BERNSTEIN POLYNOMIALS

Weakly singular integral equations are of crucial importance and have been discussed widely in the literature. There exist a number of powerful methods for solving integral equations of such kinds approximately such as Nystorm and Galerkin methods. Bhattacharya and Mandal (2010) employed expansion in terms of Bernstein polynomials to obtain approximate numerical solutions of some weakly singular integral equations with logarithmic singularities in their kernels. Capobianco and Mastronardi

(1998) employed the interpolation method to obtain an approximate solution to a Volterra type integral equation with constant coefficients containing a logarithmic kernel, after transforming the integral equation into an equivalent singular integral equation with Cauchy type kernel. An approximate numerical solution was obtained by using the invariance property of orthogonal polynomials for Cauchy singular integral equations. Khater et al (2008) used the method of finite Legendre expansion to solve Volterra integral equations with logarithmic kernels. Maleknejad et al (2007) used Galerkin wavelet method to solve certain logarithmically singular first kind integral equations. Here a straightforward method based on truncated expansion of the unknown function involving Bernstein polynomials is employed for these integral equations for the purpose of obtaining their approximate solutions.

(a) *Fredholm integral equation of second kind*

We consider a general Fredholm integral equation of the second kind with logarithmic kernel given by,

$$\alpha \ \varphi(x) + \beta \int_a^b ln \ |x - t| \ \varphi(t) \ dt = f(x), \quad a \le x \le b \quad (8.5.1)$$

where α and β are arbitrary constants. To find an approximate solution of (8.5.1), we approximate the unknown function $\varphi(x)$ using the Bernstein polynomials $B_{i,n}(x)$ defined in the interval $[a,b]$ by

$$B_{i,n}(x) = \binom{n}{i} \frac{(x-a)^i \ (b-x)^{n-i}}{(b-a)^n}, \quad i = 0,1,2,...,n, \quad (8.5.2)$$

using an expansion of the form

$$\varphi(x) \approx \sum_{i=0}^n a_i \ B_{i,n}(x), \quad a \le x \le b \quad (8.5.3)$$

where a_i's $(i = 0,1,...,n)$ are unknown constants to be determined. Substituting (8.5.3) in (8.5.1) we obtain

$$\sum_{i=0}^n a_i \left[\alpha \ B_{i,n}(x) + \beta \int_a^b ln \ |x - t| \ B_{i,n}(t) \ dt \right] = f(x), \quad a \le x \le b. \quad (8.5.4)$$

Putting $x = x_j$ $(j = 0,1,...,n)$, x_j's being chosen to be a set of suitable distinct points in (a,b), we obtain the linear system

$$\sum_{i=0}^{n} a_i A_{ij} = f_j, \; j = 0,1,...,n \tag{8.5.5}$$

where

$$A_{ij} = \left[\alpha \, B_{i,n}(x_j) + \beta \int_a^b \ln \left| x_j - t \right| B_{i,n}(t) \, dt \right], \; i,j = 0,1,...,n \tag{8.5.6}$$

and

$$f_j = f(x_j), \;\; j = 0,1,...,n. \tag{8.5.7}$$

It may be noted that A_{ij}'s and f_j's can easily be computed numerically. The linear system (8.5.5) is now solved to obtain the unknown constants a_i $(i = 0,1,...,n)$. These are then used in (8.5.3) to obtain the unknown function approximately.

For $\alpha = 0$ and $\beta = 1$, we have the first kind Fredholm integral equation given by

$$\int_a^b \ln \left| x - t \right| \varphi(t) \, dt = f(x), \;\; a \le x \le b. \tag{8.5.8}$$

In the linear system (8.5.6) A_{ij}'s are given by

$$A_{ij} = \int_a^b \ln \left| x\text{-}t \right| B_{i,n}(t) dt, i, j = 0,1,...,n. \tag{8.5.9}$$

(b) *Volterra integral equation of second kind*

Consider a general Volterra integral equation given by

$$\alpha \, \varphi(x) + \beta \int_a^x \ln \left| x - t \right| \varphi(t) \, dt = f(x), \;\; a \le x \le b \tag{8.5.10}$$

α, β being arbitrary constants. Here again, the unknown function $\varphi(x)$ is approximated in the interval $[a,b]$ as

$$\varphi(x) \approx \sum_{i=0}^{n} c_i \, B_{i,n}(x), \;\; a \le x \le b. \tag{8.5.11}$$

Again, by suitable choice of $x = x_j$, $j = 0,1,...,n$, the equation (8.5.10) can be converted into the linear system

$$\sum_{i=1}^{n} c_i \, D_{ij} = f_j, \quad j = 0,1,...,n \tag{8.5.12}$$

where

$$D_{ij} = \alpha \, B_{i,n}(x_j) + \beta \int_{a}^{x_j} \ln |x_j - t| \, B_{i,n}(t) \, dt, \quad i, j = 0,1,...,n \tag{8.5.13}$$

and

$$f_j = f(x_j), \quad j = 0,1,...,n. \tag{8.5.14}$$

The system (8.5.12) is solved for the unknown constants c_i, $i = 0,1,...,n$.

Illustrative Examples

We now illustrate the above method for a Volterra-type integral equation and two first kind Freedholm integral equations with logarithmic kernels.

Example 1: Consider the Volterra-type integral equation given by

$$a \int_{-1}^{x} \varphi(t) \, dt + \frac{b}{\pi} \int_{-1}^{1} \varphi(t) \ln |t - x| \, dt = w(x) \, x(1 - x^2), \quad -1 \le x \le 1, \tag{8.5.15}$$

where

$$w(x) = (1 - x)^{1/4} (1 + x)^{3/4} \tag{8.5.16}$$

and $a^2 + b^2 = 1$.

A direct differentiation of (8.5.15) with respect to x reduces it to a Carleman singular integral equation with Cauchy type kernel. It was shown by Capobianco and Mastronardi (1998) that a direct analytical solution of (8.5.16) is given by

$$\varphi(x) = a \left(-4x^2 + \frac{1}{2}x + 1 \right) w(x) + \frac{b}{\pi} \left[\left(\frac{1}{2} - x \right) \left\{ x \ln \frac{1-x}{1+x} + \frac{4 - 6x^2}{3(1 - x^2)} \right\} \right. \tag{8.5.17}$$

$$\left. - 6x + (1 - 3x^2) \ln \frac{1-x}{1+x} \right] w(x).$$

Now substituting

$$\varphi(x) \approx \sum_{i=0}^{n} a_i \, B_{i,n}(x), \quad -1 \le x \le 1 \qquad (8.5.18)$$

in (8.5.15) and putting $\quad x = x_j, j = 0,1,...,n \quad$ where $\quad x_j = -1 + \dfrac{2}{n+1}j,$ we obtain the linear system

$$\sum_{i=0}^{n} a_i \, D_{ij} = f_j, \quad j = 0,1,...,n \qquad (8.5.19)$$

where

$$
D_{ij} = \frac{1}{2^{n+\frac{1}{2}}} \binom{n}{i} \sum_{k=0}^{i} \sum_{l=0}^{n-i} \binom{i}{k}\binom{n-i}{l}(-1)^l \left[\frac{x_j^{k+l+1} + (-1)^{k+l}}{k+l+1} \right.
$$

$$
+ \frac{1}{\pi} \sum_{r=0}^{k+l} \binom{k+l}{r} x^{k+l-r} \left\{ (-1)^r \frac{(1+x_j)^{r+1}}{r+1} \left(\ln\,(1+x_j) - \frac{1}{r+1} \right) \right. \qquad (8.5.20)
$$

$$
\left. \left. + \frac{(1-x_j)^{r+1}}{r+1} \left(\ln\,(1-x_j) - \frac{1}{r+1} \right) \right\} \right], i, j = 0,1,...,n.
$$

and

$$f_j = w(x_j)\, x_j \left(1 - x_j^2\right), \quad j = 0,1,...,n. \qquad (8.5.21)$$

Choosing $a = b = \dfrac{1}{2^{1/2}}$ and $n = 27,$ the linear system (8.5.19) is solved for the unknown constants $a_i \ (i = 0,1,...,27).$ The following Table 1 gives a comparison between the values of the function $\varphi(x)$ obtained by the present method and that of the exact value given by (8.5.17) at different points. From this table it is clear that the method gives a good approximation.

Table 1

x	-0.8	-0.6	-0.4	-0.2	0.2	0.4	0.6	0.8
$\varphi(x)$ (approx)	-0.4191	-0.1726	0.6947	0.9509	0.4387	-0.2432	-0.9887	-1.3475
$\varphi(x)$ (exact)	-0.4266	0.1660	0.6883	0.9443	0.4304	-0.2536	-1.0037	-1.3894

Example 2: Consider the first kind Fredhokm integral equation

$$\int_0^1 ln \, |x-t| \, \varphi(t) \, dt = -2\pi (2x-1), \quad 0 \le x \le 1 \qquad (8.5.22)$$

whose exact solution is (cf. Maleknejad et al (2007))

$$\varphi(x) = \frac{4(2x-1)}{\left\{1-(2x-1)^2\right\}^{1/2}}. \qquad (8.5.23)$$

Approximating $\varphi(t)$ on $[0,1]$ using the Bernstein polynomials and following the method illustrated above, we obtain the linear system

$$\sum_{i=0}^n a_i \, A_{ij} = f_j, \quad j = 0,1,...,n \qquad (8.5.24)$$

where

$$A_{ij} = \binom{n}{i} \sum_{k=0}^{n-i} \sum_{l=0}^{i+k} \binom{n-i}{k}\binom{i+k}{l}(-1)^k \frac{x_j^{i+k-l}}{l+1}\left\{(-1)^l x_j^{l+1}\left(ln \, x_j - \frac{1}{l+1}\right)\right.$$
$$\left. +(1-x_j)^{l+1}\left(ln \, (1-x_j) - \frac{1}{l+1}\right)\right\}, \quad i,j = 0,1,...,n \qquad (8.5.25)$$

and

$$f_j = -2\pi \left(2x_j - 1\right), \quad j = 0,1,...,n. \qquad (8.5.26)$$

Choosing $n = 17$ and $x_j = \dfrac{j}{n+1}$, the linear system (8.5.24) is solved and the unknown function $\varphi(x)$ is obtained approximately. The Table 2 gives a comparison between the results of the present method and the exact values obtained from (8.5.23) at $x = 0.1, 0.2, ..., 0.9$.

Table 2

x	0.1	0.2	0.3	0.4	0.5	0.6	0.7	0.8	0.9
$\varphi(x)$ (approx.)	-5.0266	-2.9160	-1.7118	-0.8033	0.0000	0.8022	1.7100	2.9122	5.0136
$\varphi(x)$ (exact)	-5.3333	-3.0000	-1.1746	-0.8165	0.0000	0.8165	1.7457	3.0000	5.3333

Example 3: Consider the first kind Fredholm integral equation

$$\int_0^1 \ln |x-t| \, \varphi(t) \, dt = \frac{1}{2}\left[x^2 \ln x + \left(1-x^2\right) \ln (1-x) - x - \frac{1}{2}\right], \quad 0 \leq x \leq 1. \quad (8.5.27)$$

whose exact solution is (cf. Maleknejad et al (2007))

$$\varphi(x) = x. \quad (8.5.28)$$

Following the method illustrated above, we obtain the linear system

$$\sum_{i=0}^n a_i \, A_{ij} = d_j, \quad j = 0,1,...,n \quad (8.5.29)$$

where A_{ij}'s are given by (8.5.25) and

$$d_j = \frac{1}{2}\left[x_j^2 \ln x_j + \left(1-x_j^2\right) \ln (1-x_j) - x_j - \frac{1}{2}\right], \quad 0 \leq x \leq 1. \quad (8.5.30)$$

Choosing $n = 6$ and $x_j = \frac{j}{7}$, $0,1,...,6$, the system (8.5.29) is solved, and a comparison between the approximate and exact results is given in Table 3. The table 3 shows that the results are exactly same.

Table 3

x	0.1	0.2	0.3	0.4	0.5	0.6	0.7	0.8	0.9
$\varphi(x)$ (approx)	0.1	0.2	0.3	0.4	0.5	0.6	0.7	0.8	0.9
$\varphi(x)$ (exact)	0.1	0.2	0.3	0.4	0.5	0.6	0.7	0.8	0.9

Convergence Analysis

The Fredholm integral equation (8.5.1) can be expressed in the operator form

$$\left((I+K)\varphi\right)(x) = f(x), \quad a \leq x \leq b \quad (8.5.31)$$

where I is the identity operator and $(K\varphi)(x)$ denotes

$$(K\varphi)(x) = \int_a^b \ln |x-t| \, \varphi(t) \, dt. \quad (8.5.32)$$

The norm $\|\varphi\|$ is defined as

$$\|\varphi\| = \sup_{a \leq t \leq b} |\varphi(t)|.$$

For the integral equation (8.5.31), $\varphi(t)$ is approximated in terms of Bernstein polynomials as

$$\varphi(t) \simeq \varphi_n(t) = \sum_{i=0}^{n} a_i B_{i,n}(t).$$

For determining the unknown constants a_i $(i = 0,1,...,n)$ the collocation points x_j $(j = 0,1,...,n)$ used in the linear system (8.5.5) are chosen as

$$x_j = a + \frac{b-a}{n+1} j \ (j = 0,1,...,n).$$

The convergence of the polynomial $\varphi_n(t)$ to $\varphi(t)$ can be obtained by following the arguments given by Vainikko and Uba (1983) and it is found that if $f \in C^m[a,b]$, then

$$\|\varphi - \varphi_n\| < \frac{1}{n^\mu}, \ 1 \leq \mu \leq m. \tag{8.5.33}$$

Then the present method when applied to a Fredholm integral equation with a logarithmically singular kernel provides a convergent numerical method.

The Volterra type integral equation (8.5.15), as mentioned earlier, was solved by Capobianco and Mastronardi (1998) explicitly by reducing it to a Carleman singular integral equation by simple differentiation of the two sides. This is also discussed in the treatise by Kythe and Puri (2002). The Carleman equation possesses an explicit solution. Capobianco and Mastronardi (1998) used this explicit solution to obtain an approximate solution by considering the interpolation of the known function on the right of (8.5.15) and substituting this in the solution. Using the invariance properties of orthogonal polynomials for the Cauchy integral equation, they obtained the approximate solution and also gave weighted norm estimate for the error. It is now well known that the Carleman singular integral equation when approximated by Chebyshev polynomials, has the error estimate cn^{-r} when $f \in C^n[-1,1]$ (cf. Golberg and Chen (1997, Chap.7). Since the equation (8.5.15) reduces to a Carleman singular integral equation, it may be assumed that when $\varphi(x)$ satisfying (8.5.15) is approximated by Chebyshev polynomials, the error estimate will be similar. Again, as Bernstein polynomials can be expressed in terms of Chebyshev polynomials, the present method of solving the Volterra type integral equation of the form (8.5.15) yields a similar estimate.

For $\alpha = 0$ and $\beta = 0$, the integral equation (8.5.1) becomes a first kind integral equation with logarithmically singular kernel. This is known as Symm's integral equation in the literature. Sloan and Stephen (1992) studied this integral equation for its numerical solution by expanding the unknown function in terms of truncated series of Chebyshev polynomials of first kind multiplied by an appropriate weight function, and employed a collocation method to obtain the unknown constants of the truncated series. If the function f on the right side is smooth, it was shown by Sloan and Stephen (1992) that this process yields *faster-than-polynomial* convergence. In the examples 2 and 3 above, Bernstein polynomials have been used in the expansion in place of the Chebyshev polynomials. However, Bernstein polynomials can be expressed in terms of Chebyshev polynomials, and thus the present method also yields the same type of convergence. This is reflected in the illustrative examples 2 and 3.

8.6 NUMERICAL SOLUTION OF AN INTEGRAL EQUATION OF SOME SPECIAL TYPE

The problem of determining the crack energy and distribution of stress in the vicinity of a cruciform crack leads to the formulation of the integral equation of the second kind.

$$\varphi(x) + \int_0^1 K(x,t)\, \varphi(t)\, dt = f(x), \quad 0 < x \le 1 \qquad (8.6.1)$$

where

$$K(x,t) = \frac{4}{\pi} \frac{xt^2}{\left(x^2 + t^2\right)^{1/2}}, \quad 0 < x, t \le 1. \qquad (8.6.2)$$

The derivation of the integral equation (8.6.1) can be found in the works of Rooke and Sneddon (1969) and Stallybrass (1970). Here $f(x)$ is a known function depending on the internal pressure. As the value of $\varphi(x)$ at the point $x = 1$ directly relates to the stress intensity factor at the crack tips, evaluation of $\varphi(1)$ is important. Also the kernel $K(x,t)$ has a singularity at the point $(0,0)$, which results in computational complexities.

Tang and Li (2007) solved (8.6.1) using a Taylor series expansion and compared the values of $\varphi(t)$ with those of the values obtained by

Stallybrass (1970), who employed Wiener Hopf technique to solve (8.6.1). Rooke and Sneddon (1969) also gave another solution to (8.6.1) where they used Fourier Legendre series to reduce the integral equation to that of an infinite system of simultaneous linear equations. However, the methods of Stallybrass (1970) and Rooke and Sneddon (1969) are somewhat elaborate, while the method of Tang and Li (2007) makes use of Cramer's rule where calculations become tedious as terms in the approximations increase. Also Elliot (1997) used various sigmoidal transformations to find approximate solution of the integral equation (8.6.1), for the special case when $f(x) = 1$. Bhattacharya and Mandal (2010) used two simple methods, one based on approximation in terms of the Bernstein polynomials, and the other based on the rationalized Haar functions, to obtain numerical solution of (8.6.1). These are now discussed below.

The Bernstein polynomial method

To find an approximate solution of the integral equation (8.6.1), $\varphi(x)$ is approximated using the Bernstein polynomial basis in $[0,1]$ as

$$\varphi(x) = \sum_{i=0}^{n} a_i \, B_{i,n}(x) \tag{8.6.3}$$

where $B_{i,n}(x)$ $(i = 0,1,...,n)$ are the Bernstein polynomials of degree n defined on $[0,1]$ as

$$B_{i,n}(x) = \binom{n}{i} x^i (1-x)^{n-i}, \quad i = 0,1,...,n, \tag{8.6.4}$$

and a_i $(i = 0,1,...,n)$ are unknown constants to be determined. Substituting (8.6.3) in (8.6.1) we get

$$\sum_{i=0}^{n} a_i \left[B_{i,n}(x) + \int_{0}^{1} K(x,t) \, B_{i,n}(t) \, dt \right] = f(x), \quad 0 < x \le 1. \tag{8.6.5}$$

Multiplying both sides of (8.6.5) by $B_{j,n}(x)$ $(j = 0,1,...,n)$ and integrating with respect to x from 0 to 1 we obtain the linear system

$$\sum_{i=0}^{n} a_i \, C_{ij} = f_j, \quad j = 0,1,...,n \tag{8.6.6}$$

where

$$C_{ij} = D_{ij} + E_{ij} \tag{8.6.7}$$

with

$$D_{ij} = \int_0^1 B_{i,n}(x)\, B_{j,n}(x)\, dx$$

$$= \binom{n}{i} \frac{\Gamma(i+j+1)\,\Gamma(2n-i-j+1)}{\Gamma(2n+2)} \tag{8.6.8}$$

and

$$E_{ij} = \frac{4}{\pi} \int_0^1 \left\{ \int_0^1 \frac{xt^2}{(x^2+t^2)}\, B_{i,n}(t)\,dt \right\} B_{j,n}(x)\, dx$$

$$= \frac{4}{\pi}\binom{n}{i} \sum_{m=0}^{n-i} \sum_{l=0}^{n-j} \binom{n-i}{m}\binom{n-j}{l}(-1)^{l+m} \frac{1}{8+j+l+i+m}\left[(j+l)\left\{ \Psi\left(\frac{2+j+l}{4}\right) \right.\right. \tag{8.6.9}$$

$$\left.\left. - \Psi\left(\frac{4+j+l}{4}\right)\right\} + (1+i+m)\left\{ \Psi\left(\frac{1-i-m}{4}\right) - \Psi\left(\frac{3-i-m}{4}\right)\right\}\right.$$

$$\left. - 2\pi\,(-1)^{1+i+m}\,(1+i+m)\sec\left(\frac{(i+m)\pi}{2}\right)\right],$$

$\Psi(z)$ being the logarithmic derivative of the gamma function, given by

$$\Psi(z) = \frac{\Gamma'(z)}{\Gamma(z)}, \tag{8.6.10}$$

and

$$f_j = \int_0^1 f(x)\, B_{j,n}(x)\, dx. \tag{8.6.11}$$

If the internal pressure applied to each arm of the crack is chosen to be $x^{\rho-1}$, $0 \le x < 1$, ρ being an integer, then $f(x)$ is given by

$$f(x) = \frac{\Gamma\left(\dfrac{\rho}{2}\right) x^{\rho-1}}{\pi^{1/2}\, \Gamma\left(\dfrac{\rho+1}{2}\right)}. \tag{8.6.12}$$

Then for this case

$$f_j = \frac{\Gamma\left(\dfrac{\rho}{2}\right)\Gamma(j-\rho-1)\,\Gamma(n-j+1)}{\pi^{1/2}\,\Gamma\left(\dfrac{\rho+1}{2}\right)\Gamma(1+n+\rho)}. \tag{8.6.13}$$

The linear system (8.6.6) is now solved for the unknowns a_i $(i = 0,1,...,n)$ by standard numerical method to obtain $\varphi(x)$ approximately. Different values of $\varphi(1)$, which gives the stress intensity factor in the cruciform crack, are obtained for $\rho = 1, 2, ..., 10$. A comparison between the exact results as given in Tang and Li (2007) and the results obtained by the present method for $n = 12$ is given in Table 1. This table shows that the results are fairly accurate.

The rationalized Haar function method

The rationalized Haar functions have been used to solve a number of Fredholm integral equations by Maleknejad and Mirazee (2003,2005) and some Volterra integral equations by Reihani and Abadi (2007) with regular kernels. This method is applied here to solve (8.6.1) whose kernel has a singularity at one end. It may be noted that the idea of using Haar functions come from the rapid convergence feature of Haar series in the expansion of functions. It is thus very useful to approximate the unknown function in various problems of mathematical physics, which can be effectively reduced to differential and integral equations. Calculations involved in the solution process by this method are very simple and requires merely some matrix operations.

Table 1: Numerical values of $\varphi(1)$

ρ	Exact results	Approximate results
1	0.86354	0.86352
2	0.57547	0.57535
3	0.46350	0.46349
4	0.39961	0.39963
5	0.35681	0.35682
6	0.32549	0.32549
7	0.30125	0.30125
8	0.28176	0.28176
9	0.26564	0.26564
10	0.25201	0.25201

The orthogonal set of Haar functions considered here are a group of square waves with magnitudes $2^{j/2}, 2^{-j/2}$ and 0, $j = 0,1,....$ Further, the Haar functions are the rationalized Haar functions obtained by deleting the irrational numbers and introducing integer powers of 2. Thus the rationalized Haar functions, which retain all the properties of the original Haar functions, have magnitudes $1, -1$ and 0.

The rationalized Haar functions are defined as

$$\Phi_m(x) = \begin{cases} 1, & J_1 \le x < J_{1/2}, \\ -1, & J_{1/2} \le x < J_0, \\ 0, & \text{otherwise} \end{cases} \qquad (8.6.14)$$

where

$$J_\mu = \frac{j-\mu}{2^i}, \quad \mu = 0, \frac{1}{2}, 1. \qquad (8.6.15)$$

The values of m is given in terms of i and j as

$$m = 2^i + j - 1, \quad i = 0, 1, \ldots, j = 1, 2^1, 2^2, \ldots, 2^i \qquad (8.6.16)$$

and

$$\Phi_0(x) = 1, \quad 0 \le x < 1, \quad i = 0, j = 0. \qquad (8.6.17)$$

Also, the orthogonal property of Haar functions gives

$$\int_0^1 \Phi_m(x)\, \Phi_l(x)\, dx = \begin{cases} 2^{-i}, & m = l, \\ 0, & \text{otherwise.} \end{cases} \qquad (8.6.18)$$

Now, any function $g(x)$, which is square integrable over $[0,1]$, can be expanded using the rationalized Haar functions as

$$g(x) = \sum_{r=0}^{\infty} d_r\, \Phi_r(x) \qquad (8.6.19)$$

so that

$$d_r = 2^i \int_0^1 g(x)\, \Phi_r(x)\, dx, \quad r = 0, 1, \ldots. \qquad (8.6.20)$$

Therefore, to find an approximate solution of the integral equation (8.6.1), we approximate $\varphi(x)$ using the rationalized Haar functions as

$$\varphi(x) \simeq \sum_{r=0}^{p-1} d_r\, \Phi_r(x) \qquad (8.6.21)$$

$$= d^T \Phi(x) \qquad (8.6.22)$$

with

$$p = 2^i + j - 1, \quad i = 0, 1, ..., \alpha; \, j = 1, 2, ..., 2^i,$$

and

$$\boldsymbol{d} = \left[d_0, d_1, ..., d_{p-1} \right]^T \tag{8.6.23}$$

$$\Phi(x) = \left[\Phi_0(x), \Phi_1(x), ..., \Phi_{p-1}(x) \right]^T \tag{8.6.24}$$

The kernel $K(x,t)$ is also approximated by using the rationalized Haar functions as

$$K(x,t) = \sum_{u=0}^{p-1} \sum_{v=0}^{p-1} h_{uv} \Phi_u(x) \Phi_v(t) \tag{8.6.25}$$

$$= \Phi^T(x) \ \boldsymbol{H} \Phi(t) \tag{8.2.26}$$

where

$$h_{uv} = 2^{\mu + \nu} \int_0^1 \int_0^1 K(x,t) \Phi_u(x) \Phi_v(t) dx \, dt, \tag{8.6.27}$$

$$\boldsymbol{H} = \left[h_{uv} \right]_{p \times p}, \quad u = 2^\mu + \gamma - 1, \ 2^\nu + \delta - 1;$$

$$\mu, \nu = 0, 1, ..., \alpha; \ \gamma = 1, 2, ..., 2^\mu; \ \delta = 1, 2..., 2^\nu.$$

By truncating (8.6.19) upto a finite number of terms, it can be shown that

$$\left[g\left(\frac{1}{2p}\right), g\left(\frac{3}{2p}\right), ..., g\left(\frac{2p-1}{2p}\right) \right] = \boldsymbol{d}^T \hat{\Phi} \tag{8.6.28}$$

with

$$\hat{\Phi} = \left[\Phi\left(\frac{1}{2p}\right), \ \Phi\left(\frac{3}{2p}\right), ..., \Phi\left(\frac{2p-1}{2p}\right) \right]. \tag{8.6.29}$$

Hence

$$\boldsymbol{H} = \left(\hat{\Phi}^{-1} \right)^T \ \hat{\boldsymbol{H}} \hat{\Phi}^{-1} \tag{8.6.30}$$

where

$$\boldsymbol{H} = \left[K\left(\frac{2m-1}{2p}, \frac{2n-1}{2p}\right) \right]_{p \times p}, \quad m, n = 1, 2, ..., p. \tag{8.6.31}$$

Similarly $f(x)$ in (8.6.1) can be approximated using $\Phi_r(x)$ as

$$f(r) \simeq \sum_{r=0}^{p-1} f_r \, \Phi_r(x) \qquad (8.6.32)$$

$$= F^T \Phi(x)$$

with

$$F = \left[f_0, f_1, ..., f_{p-1} \right] = \hat{\Phi}^{-1} \hat{F}^T \qquad (8.6.33)$$

where

$$\hat{F} = \left[f\left(\frac{2n-1}{2p} \right) \right]_{1 \times p} , n = 1, 2, ..., p. \qquad (8.6.34)$$

Using these evaluations, the equation (8.6.1) produces

$$d^T \Phi + \int_0^1 \Phi^T(x) \, ^{H\Phi}(t) \, d^T \Phi(t) \, dt = F^T \Phi(x) \qquad (8.6.35)$$

which reduces to

$$\left[I_p + H \int_0^1 \Phi(t) \, \Phi^T(t) dt \right] d = F, \qquad (8.6.36)$$

I_p being the identity matrix of order p, and

$$\int_0^1 \Phi(t) \, \Phi^T(t) dt = D, \qquad (8.6.37)$$

D being the diagonal matrix given by

$$D = diag \left(1; 1; \frac{1}{2}, \frac{1}{2} \underbrace{\frac{1}{2^2}, \frac{1}{2^2}, \frac{1}{2^3}, \frac{1}{2^3}}_{2^2} \underbrace{\frac{1}{2^3}, ..., \frac{1}{2^3}}_{2^3}, ..., \underbrace{\frac{1}{2^\alpha}, ..., \frac{1}{2^\alpha}}_{2^\alpha} \right) \qquad (8.6.38)$$

Therefore (8.6.35) further reduces to

$$\left(I_p + HD \right) d = F \qquad (8.6.39)$$

so that

$$d = \left(I_p + HD\right)^{-1} F. \tag{8.6.40}$$

Thus the unknown constants d_r $(r = 0,1,...,p - 1$ are obtained and these then given an approximation to $\varphi(x)$ after using (8.6.21). From this, $\varphi(t)$ is obtained. Table 2 shows comparison between the exact values of $\varphi(1)$ as given in Tang and Li (2007) to those obtained by the present method for $\alpha = 4, 5$ and 6. From this table it is observed that the results are fairly accurate and improve as α (i.e. p) increases.

Table 2: Numerical values of $\varphi(t)$ for different values of α

ρ	Exact results	Approximate results		
		$\alpha = 4, p = 3.2$	$\alpha = 5, p = 64$	$\alpha = 6, p = 128$
1	0.86354	0.98307	0.99995	0.99998
2	0.57547	0.62659	0.63162	0.63412
3	0.46350	0.48445	0.49220	0.49609
4	0.39961	0.40479	0.41453	0.41945
5	0.35681	0.35208	0.36341	0.36917
6	0.32549	0.31380	0.32646	0.33295
7	0.30125	0.28431	0.29813	0.30524
8	0.28176	0.26063	0.27547	0.28316
9	0.26564	0.24106	0.25680	0.26500
10	0.25201	0.22449	0.24105	0.24973

8.7 NUMERICAL SOLUTION OF A SYSTEM OF GENERALIZED ABEL INTEGRAL EQUATIONS

Here a system of generalized Abel integral equations arising in certain mixed boundary value problems in the classical theory of elasticity, is considered. Pandey and Mandal (2010) solved this system numerically by using Bernstein polynomials. This is now described briefly.

Many interesting problems of mechanics and physics lead to an integral equation in which the kernel $K(t, u)$ is of convolution type, that is $K(t, u) = k(t - u)$, where $k(x)$ is a certain function of one variable. A similar integral equation with weakly singular kernel, named as Abel integral equation, appears in many branches of science and engineering. Usually, physical quantities accessible to measurement are quite often related to physically important but experimentally inaccessible ones by Abel integral equations. A system of generalized integral equations of

Abel type were studied by Lowengrub(1976) and Walton(1979). As stated by Walton (1979), certain mixed boundary value problems arising in the classical theory of elasticity reduce to the problem of determining functions ϕ_1 and ϕ_2 satisfying Abel type integral equations of the type,

$$\left. \begin{array}{l} a_{11}(x)\displaystyle\int_0^x \frac{\phi_1(t)}{(x^\alpha - t^\alpha)^\mu}\,dt + a_{12}(x)\displaystyle\int_x^1 \frac{\phi_2(t)}{(t^\alpha - x^\alpha)^\mu}\,dt = f_1(x) \\[4mm] a_{21}(x)\displaystyle\int_x^1 \frac{\phi_1(t)}{(t^\alpha - x^\alpha)^\mu}\,dt + a_{22}(x)\displaystyle\int_0^x \frac{\phi_2(t)}{(x^\alpha - t^\alpha)^\mu}\,dt = f_2(x) \end{array} \right\} \quad (8.7.1)$$

where, $x \in (0,1)$, $0 < \mu < 1$, $\alpha \geq 1$ and $a_j(x)$ are continuous on $[0,1]$. Since only $\alpha = 1, 2$ occur in physical problems, we restrict our attention to these values only. Mandal et al (1996), solved this problem analytically for the special case $\alpha = 1$, $\mu = 1/2$, $a_j(x) = 1$, using the idea of fractional calculus. Here a simple algorithm based on approximation of unknown functions in terms of Bernstein polynomials has been used to find numerical solutions by converting and solving the generalized system into a system of linear equations for given $f_i(x)$ $(i = 1, 2)$.

For obtaining the approximate solution of (8.7.1), $\phi_1(t)$ and $\phi_2(t)$ are approximated in the Bernstein polynomial basis on $[0,1]$ as,

$$\phi_1(t) \approx \widetilde{\phi}_1(t) = \sum_{i=0}^n a_i B_{i,n}(t) \quad \text{and} \quad \phi_2(t) \approx \widetilde{\phi}_2(t) = \sum_{i=0}^n b_i B_{i,n}(t) \quad (8.7.2)$$

where a_i, and b_i $(i = 0,1,2,\ldots,n)$ are unknown constants to be determined. Substituting (8.7.2) in (8.7.1), we obtain,

$$a_{11}(x)\sum_{i=0}^n a_i \lambda_i(x) + a_{12}(x)\sum_{i=0}^n b_i \beta_i(x) = f_1(x)$$

$$a_{21}(x)\sum_{i=0}^n a_i \beta_i(x) + a_{22}(x)\sum_{i=0}^n b_i \lambda_i(x) = f_2(x) \qquad (8.7.3)$$

where,

$$\lambda_i(x) = \int_0^x \frac{B_{i,n}(t)}{(x^\alpha - t^\alpha)^\mu}\,dt, \quad \beta_i(x) = \int_x^1 \frac{B_{i,n}(t)}{(t^\alpha - x^\alpha)^\mu}\,dt \quad \text{and} \quad x \in (0,1).$$

We now put $x = x_j, j = 0,1,...,n$ in (8.7.3), where x_j's are chosen as suitable distinct points in $(0,1)$ such that $0 < x_0 ... < x_n < 1$. Putting $x = x_j$ we obtain the linear system

$$a_{11}(x_j) \sum_{i=0}^{n} a_i \lambda_{ij} + a_{12}(x_j) \sum_{i=0}^{n} b_i \beta_{ij} = f_1(x_j)$$

(8.7.4)

$$a_{21}(x_j) \sum_{i=0}^{n} a_i \beta_{ij} + a_{22}(x_j) \sum_{i=0}^{n} b_i \lambda_{ij} = f_2(x_j)$$

where,

$$\lambda_{ij} = \lambda_i(x_j) \text{ and } \beta_{ij} = \beta_i(x_j), \text{ for } i, j = 0,1,...,n.$$

The linear system (8.7.4) can be easily solved by any standard method for the unknown constants a_i's and b_i's provided of course the coefficient matrix is nonsingular. It is emphasized that it is always possible to choose distinct points $x_j \in (0,1)$ ($j = 0,1,...,n$) such that this is possible. The computed $\underset{\sim}{a_i}$'s and $\underset{\sim}{b_i}$'s are then used in (8.7.2) to obtain the approximate solutions $\tilde{\phi}_1(t)$ and $\tilde{\phi}_2(t)$.

A number of illustrative examples are now presented to demonstrate the simplicity of the method as well as accuracy of the numerical results. In all these examples we have chosen $n = 8$ and the errors, defined as $E_1(t) = \tilde{\phi}_1(t) - \phi_1(t)$ and $E_2(t) = \tilde{\phi}_2(t) - \phi_2(t)$, are computed for different values of $t \in [0,1]$ and depicted graphically. Also we tabulated the approximate and exact solution for $t = 0.0, 0.2, ..., 0.8, 1.0$.

Illustrative examples

Case1: In this case, we have considered examples with the assumption that $\alpha = 1$.

Example1.1 Consider the system of generalized Abel integral

equations (8.7.1) with $a_{ij}(x) = 1$, for $i, j = 1,2$, $\mu = 1/2$ with

$$f_1(x) = \frac{4}{3}x^{3/2} + \frac{2}{5}(1-x)^{5/2} + 2x^2(1-x)^{1/2} + \frac{4}{3}x(1-x)^{3/2},$$

$$f_2(x) = \frac{2}{3}(1-x)^{3/2} + 2x(1-x)^{1/2} + \frac{16}{15}x^{5/2}.$$

This has the exact solution $\phi_1(t) = t$ and $\phi_2(t) = t^2$.

Table 1: Approximate and exact solution of example 1.1

t	0.0	0.2	0.4	0.6	0.8	1.0
$\tilde{\phi}_1(t)$	-0.000110	0.200022	0.400040	0.600035	0.800067	0.998009
$\phi_1(t)$	0.0	0.2	0.4	0.6	0.8	1.0
$\tilde{\phi}_2(t)$	-0.000063	0.040015	0.160037	0.360032	0.640060	0.998055
$\phi_2(t)$	0.0	0.04	0.16	0.36	0.64	1.0

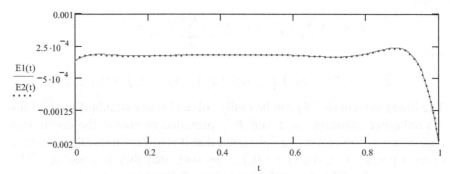

Fig. 1 Errors associated with Ex. 1.1

Example1.2 Consider the system of generalized Abel integral equation with $a_{ij}(x) =1$, for $i, j =1,2, \mu =1/3$ with

$$f_1(x)=\frac{9}{10}x^{5/3}+\frac{729}{1540}x^{14/3}+\frac{3}{11}(1-x)^{11/3}+\frac{9}{8}x(1-x)^{8/3}-\frac{3}{2}x^2(1-x)^{2/3}-\frac{6}{5}x(1-x)^{5/3},$$

$$+\frac{3}{2}x^3(1-x)^{2/3}+\frac{9}{5}x^2(1-x)^{5/3}-\frac{3}{8}(1-x)^{8/3}$$

and

$$f_2(x)=\frac{3}{5}(1-x)^{5/3}+\frac{19}{14}x^2(1-x)^{8/3}+\frac{3}{2}x^4(1-x)^{2/3}+\frac{12}{5}x^3(1-x)^{5/3}+\frac{12}{11}x(1-x)^{11/3},$$

$$+\frac{3}{2}x(1-x)^{2/3}+\frac{3}{14}(1-x)^{14/3}+\frac{243}{440}x^{11/3}-\frac{27}{40}x^{4/3}.$$

This has exact solution $\phi_1(t) =t+t^4$ and $\phi_2(t) =t^3-t^2$.

Table 2: Approximate and exact solution of example 1.2

t	0.0	0.2	0.4	0.6	0.8	1.0
$\tilde{\phi}_1(t)$	0.000147	0.201583	0.425635	0.729584	1.209559	2.001582
$\phi_1(t)$	0.0	0.201600	0.425600	0.729600	1.209600	2.0
$\tilde{\phi}_2(t)$	-0.000316	-0.032005	-0.095979	-0.143986	-0.128090	0.000498
$\phi_2(t)$	0.0	-0.032000	-0.096000	-0.144000	-0.128000	0.0

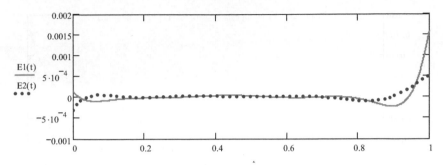

Fig. 2 Errors associated with Ex. 1.2

Example 1.3

In this example, $a_{11}(x)=1$, $a_{12}(x)=1/4$, $a_{21}(x)=1/2$, $a_{22}(x)=3$, $\mu=1/2$ with

$$f_1(x)=\frac{16}{15}x^{5/2}+\frac{1}{10}(1-x)^{5/2}+\frac{1}{2}x^2(1-x)^{3/2}+\frac{1}{2}x^2(1-x)^{1/2}+\frac{1}{2}x^3(1-x)^{1/2},$$

$$+\frac{1}{3}x(1-x)^{3/2}+\frac{1}{14}(1-x)^{7/2}+\frac{3}{10}x(1-x)^{5/2}$$

and

$$f_2(x)=\frac{1}{5}(1-x)^{5/2}+x^2(1-x)^{1/2}+\frac{2}{3}x(1-x)^{3/2}+\frac{16}{5}x^{5/2}+\frac{96}{35}x^{7/2}.$$

This has exact solution $\phi_1(t)=t^2$ and $\phi_2(t)=t^2+t^3$.

Table 3: Approximate and exact solution of example 1.3

t	0.0	0.2	0.4	0.6	0.8	1.0
$\widetilde{\phi}_1(t)$	-0.001610	0.040050	0.160083	0.359990	0.639726	1.009014
$\phi_1(t)$	0.0	0.040000	0.160000	0.360000	0.640000	1.000000
$\widetilde{\phi}_2(t)$	-0.000084	0.047977	0.223992	0.575916	1.151847	2.005230
$\phi_2(t)$	0.0	0.048000	0.224000	0.576000	1.152000	2.000000

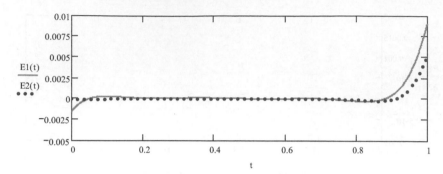

Fig. 3 Errors associated with Ex. 1.3

Example1.4 This example presents the problem (1), with
$a_{11}(x) = x^2 + 1$, $a_{12}(x) = (x+1)/4$, $a_{21}(x) = x^2/2$, $a_{22}(x) = 2 - x$,
$\mu = 1/2$ and

$$f_1(x) = \frac{16}{15}(x^2 + 1)x^{5/2} + \frac{1}{4}(1+x)\left(\frac{2}{7}(1-x)^{7/2} + 2(1-x)^{3/2} x^2 + \frac{6}{5}(1-x)^{5/2} x + 2 x^3 (1-x)^{1/2}\right)$$

$$f_2(x) = \frac{32}{35}(2-x)x^{7/2} + \frac{1}{2}x^2\left(\frac{2}{5}(1-x)^{5/2} + 2(1-x)^{1/2} x^2 + \frac{4}{3}(1-x)^{3/2} x\right).$$

This has exact solution $\phi_1(t) = t^2$ and $\phi_2(t) = t^3$.

Table 4: Approximate and exact solution of example 1.4

t	0.0	0.2	0.4	0.6	0.8	1.0
$\widetilde{\phi}_1(t)$	0.000000	0.040019	0.160049	0.359974	0.639978	1.003905
$\phi_1(t)$	0.0	0.040000	0.160000	0.360000	0.640000	1.000000
$\widetilde{\phi}_2(t)$	0.000000	0.008019	0.064023	0.215971	0.511841	1.001006
$\phi_2(t)$	0.0	0.008000	0.064000	0.216000	0.512000	1.000000

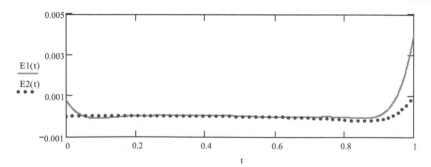

Fig. 4 Errors associated with Ex. 1.4

Example 1.5 In this example,

$$a_{11}(x)=1 \;,\; a_{12}(x)=1/4 \;,\; a_{21}(x)=2 \;,\; a_{22}(x)=1/8 \;,\; \mu=2/3 \text{ with}$$

$$f_1(x)=-\frac{243}{140}x^{10/3}+\frac{729}{455}x^{13/3}+\frac{3}{40}(1-x)^{10/3}+\frac{9}{28}x(1-x)^{7/3}-\frac{3}{4}x^2(1-x)^{1/3} \;,$$

$$-\frac{3}{8}x(1-x)^{4/3}+\frac{3}{4}x^3(1-x)^{1/3}+\frac{9}{16}x^2(1-x)^{4/3}-\frac{3}{28}(1-x)^{7/3}$$

and

$$f_2(x)=-\frac{3}{5}(1-x)^{10/3}-\frac{18}{7}x(1-x)^{7/3}+\frac{36}{7}x^2(1-x)^{7/3}+\frac{12}{5}x(1-x)^{10/3}+6x^3(1-x)^{1/3}$$

$$+6x^4(1-x)^{1/3}-\frac{9}{2}x^2(1-x)^{4/3}+\frac{6}{13}(1-x)^{13/3}+\frac{243}{1120}x^{7/3}-\frac{27}{112}x^{7/3}.$$

This has exact solution $\phi_1(t)=t^3-t^2$ and $\phi_2(t)=t^4-t^3$.

Table 5: Approximate and exact solution of example 1.5

t	0.0	0.2	0.4	0.6	0.8	1.0
$\widetilde{\phi}_1(t)$	0.000009	-0.006404	-0.038404	-0.086399	-0.102402	-0.000014
$\phi_1(t)$	0.0	-0.006400	-0.038400	-0.086400	-0.102400	0.0
$\widetilde{\phi}_2(t)$	0.000021	-0.032064	-0.095994	-0.143986	0.127987	0.000140
$\phi_2(t)$	0.0	-0.032000	-0.096000	-0.144000	0.128000	0.0

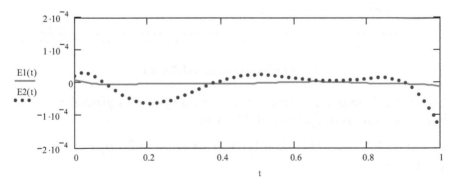

Fig. 5 Errors associated with Ex. 1.5

Case2: In this case, we have considered examples with the assumption that $\alpha = 2$.

Example 2.1 Consider the system of a generalized Abel integral equation with $a_{ij}(x) = 1$, for $i, j = 1, 2$ with

$$f_1(x) = x + \frac{1}{2}(1-x)^{1/2} + \frac{1}{2}x^2 \ln\left(1 + \sqrt{1-x^2}\right) - \frac{1}{2}x^2 \ln(x),$$

$$f_2(x) = \frac{\pi}{4}x^2 + (1-x^2)^{1/2}.$$ This has exact solution $\phi_1(t) = t$ and $\phi_2(t) = t^2$.

Table 6: Approximate and exact solution of example 2.1

t	0.0	0.2	0.4	0.6	0.8	1.0
$\tilde{\phi}_1(t)$	-0.000003	0.200011	0.400036	0.600028	0.800080	0.997813
$\phi_1(t)$	0.0	0.200000	0.400000	0.600000	0.800000	1.000000
$\tilde{\phi}_2(t)$	-0.000003	0.040016	0.160038	0.360031	0.640077	0.998037
$\phi_2(t)$	0.0	0.040000	0.160000	0.360000	0.640000	1.0

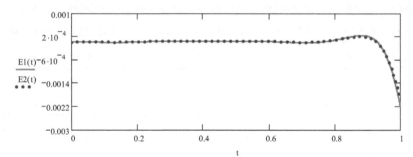

Fig. 6 Errors associated with Ex. 2.1

Example 2.2: In this example we consider a system of generalized Abel integral equations of the form (8.7.1) with

$$a_{11}(x) = x^2 + 1, \quad a_{12}(x) = (x+1)/4, \quad a_{21}(x) = x^2/2, \quad a_{22}(x) = x,$$
$$\mu = 1/2 \text{ and}$$

$$f_1(x) = (x^2 + 1)\left(\frac{3\pi}{16}x^4 + \frac{\pi}{16}x^2\right) + \left(\frac{1}{4}x + \frac{1}{4}\right)\left[\frac{1}{2}(1-x^2)^{1/2} + \frac{1}{2}x^2 \ln\left(1 + (1-x^2)^{1/2}\right) - \frac{1}{2}x^2 \ln x\right],$$

$$f_2(x) = \frac{1}{2}x^2\left(\frac{3}{4}(1-x^2)^{1/2} + \frac{1}{2}x^2 \ln\left(1 + (1-x^2)^{1/2}\right) + \frac{3}{8}x^4 \ln\left(1 + (1-x^2)^{1/2}\right) - \frac{3}{8}x^4 \ln x - \frac{1}{2}x^2 \ln x\right)$$
$$+ \frac{\pi}{4}x^3.$$

This has exact solution $\phi_1(t) = t^2 + t^4$ and $\phi_2(t) = t^2$.

Table 7: Approximate and exact solution of example 2.2

t	0.0	0.2	0.4	0.6	0.8	1.0
$\widetilde{\phi}_1(t)$	-0.000001	0.041640	0.185705	0.489598	1.049501	2.008946
$\phi_1(t)$	0.0	0.041600	0.185600	0.489600	1.049600	2.0
$\widetilde{\phi}_2(t)$	0.000000	0.040094	0.160081	0.359896	0.639494	1.003657
$\phi_2(t)$	0.0	0.040000	0.160000	0.360000	0.640000	1.0

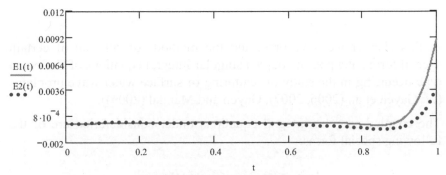

Fig. 7 Errors associated with Ex. 2.2

From the given numerical examples and tables, we conclude that the method based on approximation in terms of Berstein polynomials is a powerful tool for solving systems of generalized Abel integral equations.

Chapter 9

Some Special Types of Coupled Singular Integral Equations of Carleman Type and their Solutions

In this chapter we have presented the methods of solution of certain special forms of a pair of coupled singular integral equations of Carleman type occurring in the study of scattering of surface water wave problems (cf. Gayen et al. (2006, 2007), Gayen and Mandal (2009)).

The coupled singular integral equations under consideration are of the following special forms:

$$a(x)\,\varphi_1(x) + \frac{1}{\pi} \int_0^\infty \frac{\varphi_1(t)}{t-x}\, dt - \frac{1}{\pi} \int_0^\infty \frac{\varphi_2(t)e^{-lt}}{t+x}\, dt = f_1(x), \quad x > 0 \quad (9.1)$$

$$a(x)\,\varphi_2(x) + \frac{1}{\pi} \int_0^\infty \frac{\varphi_2(t)}{t-x}\, dt - \frac{1}{\pi} \int_0^\infty \frac{\varphi_1(t)e^{-lt}}{t+x}\, dt = f_2(x), \quad x > 0 \quad (9.2)$$

where $a(x)$, $f_1(x)$ and $f_2(x)$ are known functions, l is a known positive parameter, and φ_1 and φ_2 are unknown functions to be determined, the integrals involving $(t-x)^{-1}$ being in the sense of Cauchy principal value integrals.

The specific forms of the known function $a(x)$ which occurs in the study on water wave problems (cf. Gayen et al (2006, 2007), Gayen and Mandal (2009)) are

$$a(x) = \frac{x^2 + K_1 K_2}{x(K_1 - K_2)}, \quad K_1 > 0,\ K_2 > 0,\ K_1 \neq K_2$$

and

$$a(x) = \frac{x^2(Dx^4+1)+K^2}{DKx^5}, \quad D > 0, \ K > 0.$$

The functions $f_1(x), f_2(x)$ also have known forms.

9.1 THE CARLEMAN SINGULAR INTEGRAL EQUATION

The Carleman singular integral equation over a semi-infinite range given by

$$a(x)\ \varphi(x) + \frac{1}{\pi} \int_0^\infty \frac{\varphi(t)}{t-x}\, dx = f(x), x > 0, \qquad (9.1.1)$$

also occurs in the study of water wave problems (cf. Chakrabarti (2000)) and its solution can be obtained by reducing it to an appropriate Riemann Hilbert problem. Its method of solution is briefly described below.

Introducing a sectionally analytic function $\Phi(z)$ in the complex z-plane cut along the positive real axis as defined by

$$\Phi(z) = \frac{1}{2\pi i} \int_0^\infty \frac{\varphi(t)}{t-z}\, dt \qquad (9.1.2)$$

and utilizing the Plemelj formulae

$$\Phi^\pm(x) = \pm \frac{1}{2}\ \varphi(x) + \frac{1}{2\pi i} \int_0^\infty \frac{\varphi(t)}{t-x}\, dt, \qquad (9.1.3)$$

the equation (9.1.1) can be expressed as

$$\{a(x)+i\}\Phi^+(x) - \{a(x)-i\}\Phi^-(x) = f(x), \ x > 0 \qquad (9.1.4)$$

where $\Phi^\pm(x) = \lim_{y\to\pm0} \Phi(z), x > 0$. The relation (9.1.4) represents a Riemann Hilbert Problem for the determination of the function $\Phi(z)$.

The solution of the problem (9.1.4) can be easily obtained in the form

$$\Phi(z) = \Phi_0(z)\frac{1}{2\pi i} \int_0^\infty \frac{1}{(a(t)+i)\Phi_0^+(t)}\frac{dt}{t-z} \qquad (9.1.5)$$

where $\Phi_0(z)$ represents the solution of the homogeneous problem

$$\{a(x)+i\}\Phi_0^+(x)-\{a(x)-i\}\Phi_0^-(x)=0, \; x>0 \qquad (9.1.6)$$

and $\Phi_0^{\pm}(x) = \lim_{y\to\pm0} \Phi_0(z), \; x>0.$

$\Phi_0(z)$ is non-zero and sectionally analytic in the complex z-plane cut along the positive real axis, and is such that $|\Phi_0(z)| = 0\,(1)$ as $|z|\to\infty$. This produces $|\Phi\,(z)| = 0\left(\dfrac{1}{|z|}\right)$ as $|z|\to\infty$, as is to be expected from the representation (9.1.2) for the function $\Phi(z)$. The function $\Phi_0(z)$ is obtained as

$$\Phi_0(z) = \exp\left[\frac{1}{2\pi i}\int_0^\infty \ln\left(\frac{a(t)-i}{a(t)+i}\right)\frac{dt}{t-z}\right]. \qquad (9.1.7)$$

Thus $\Phi(z)$ is now known from (9.1.5), and the solution of the integral equation (9.1.1) is obtained by using the formula

$$\varphi(x) = \Phi^+(x) - \Phi^-(x), \; x>0. \qquad (9.1.8)$$

Remark

While the Carleman singular integral equation (9.1.1) possesses an explicit solution, the two coupled equations (9.1) and (9.2) cannot be solved explicitly due to the presence of the parameter l in the non-singular integral in each equation. Two types of methods can be employed to solve the integral equations, one of them being an approximate method valid for large values of the parameter l, and the other method casts the original integral equations into a system of Fredholm integral equations of the second kind with regular kernels which can be solved numerically for any value, large, medium or small, of the parameter l. These two methods are described in the next two sections. In both these methods, the common feature is the utility of the explicit solution of the Carleman integral equation (9.1.1).

9.2 SOLUTION OF THE COUPLED INTEGRAL EQUATIONS FOR LARGE l

The coupled integral equations (9.1) and (9.2) were solved approximately for large l by Gayen et al (2006) when $a(x), f_1(x), f_2(x)$ have the following specific forms:

$$a(x) = \frac{x^2 + K_1 K_2}{x(K_1 - K_2)}, \tag{9.2.1}$$

$$f_1(x) = \frac{\alpha}{2} \frac{e^{iK_2 l}}{x - iK_2} + \frac{\beta}{2} \frac{1}{x + iK_2}, \tag{9.2.2}$$

$$f_2(x) = \frac{\beta}{2} \frac{e^{iK_2 l}}{x - iK_2} + \frac{\alpha}{2} \frac{1}{x + iK_2}, \tag{9.2.3}$$

where $K_1 > 0$, $K_2 > 0$ and α, β are *unknown* constants. K_1, K_2 are real but α, β are in general complex. This method is described below briefly.

For very large values of l, we can ignore the integrals involving e^{-lt} in (9.1) and (9.2), and thus as zero-order approximation we obtain the following uncoupled integral equations:

$$a(x)\, \varphi_1^0(x) + \frac{1}{2} \int_0^\infty \frac{\varphi_1^0(t)}{t - x}\, dt = f_1^0(x),\ x > 0 \tag{9.2.4}$$

$$a(x)\, \varphi_2^0(x) + \frac{1}{2} \int_0^\infty \frac{\varphi_2^0(t)}{t - x}\, dt = f_2^0(x),\ x > 0 \tag{9.2.5}$$

where the superscript '0' denotes the zero-order approximations, and $f_1^0(x), f_2^0(x)$ are the same as $f_1(x), f_2(x)$ given in (9.2.2), (9.2.3) above with α, β replaced by α^0, β^0. The two independent equations (9.2.4) and (9.2.5) can be solved as in section 9.1.

The solutions are found to be

$$\varphi_1^0(x) = \alpha^0\, e^{iK_2 l}\, P_1(x) + \beta^0\, P_2(x), \tag{9.2.6}$$

$$\varphi_2^0(x) = \alpha^0\, P_2(x) + \beta^0\, P_1(x), \tag{9.2.7}$$

where

$$P_1(x), P_2(x) = \frac{1}{2}\, \frac{(K_1 - K_2)\, x \Lambda_0^+(x)}{(x - iK_1)(x - iK_2)(x \mp iK_2)\Lambda_0(\pm iK_2)} \tag{9.2.8}$$

with

$$\Lambda_0(z) = \exp\left[\frac{1}{2\pi i}\left\{\int_0^\infty \left(\ln\frac{t-iK_1}{t+iK_1} - 2\pi i\right)\frac{dt}{t-z} - \int_0^\infty\left(\ln\frac{t-iK_2}{t+iK_2} - 2\pi i\right)\frac{dt}{t-z}\right\}\right]. \quad (9.2.9)$$

The zero-order approximations α^0 and β^0 to α and β respectively can be determined from some other relations (see Gayen et al (2006) for details).

To obtain the next order (first-order) approximate solutions $\varphi_1^1(x)$, $\varphi_2^1(x)$ of the coupled equations (9.1) and (9.2), we decouple these by replacing $\varphi_2(t)$ in (9.1) by the known function $\varphi_2^0(t)$ and $\varphi_1(t)$ in (9.2) by $\varphi_1^0(t)$, and also approximate α, β appearing in $f_1(x)$ and $f_2(x)$ by α^1, β^1. This art gives rise to the following pair of Carleman singular integral equations for $\varphi_1^1(x)$, $\varphi_2^1(x)$ as

$$a(x)\,\varphi_1^1(x) + \frac{1}{\pi}\int_0^\infty \frac{\varphi_1^1(t)}{t-x}\,dt = f_1^1(x), \ x > 0, \quad (9.2.10)$$

$$a(x)\,\varphi_2^1(x) + \frac{1}{\pi}\int_0^\infty \frac{\varphi_2^1(t)}{t-x}\,dt = f_2^1(x), \ x > 0 \quad (9.2.11)$$

where

$$f_1^1(x) = \frac{1}{2}\int_0^\infty \frac{\varphi_1^0(t)e^{-lt}}{t+x}\,dt + \frac{\alpha^1}{2}\frac{e^{iK_2l}}{x-iK_2} + \frac{\beta^1}{2}\frac{1}{x+iK_2} \quad (9.2.12)$$

$$f_2^1(x) = \frac{1}{2}\int_0^\infty \frac{\varphi_2^0(t)e^{-lt}}{t+x}\,dt + \frac{\beta^1}{2}\frac{e^{iK_2l}}{x-iK_2} + \frac{\alpha^1}{2}\frac{1}{x+iK_2}. \quad (9.2.13)$$

Note that $f_1^1(x)$ and $f_2^1(x)$ contain the unknown constants α^1, β^1. (the first-order approximations to α, β).

The equations (9.2.10) and (9.2.11) can be solved as before, and it can be shown that (cf. Gayen et al (2006))

$$\varphi_1^1(x) = \alpha^1\,e^{iK_2l}\,P_1(x) + \beta^1 P_2(x) \quad (9.2.14)$$

$$\varphi_2^1(x) = \alpha^1 P_2(x) + \beta^1\,e^{iK_2l}P_1(x) \quad (9.2.15)$$

where $P_1(x), P_2(x)$ are given in (9.2.8) above. The first-order approximations α^1, β^1 to α, β can be obtained from some other relations as was in the case for zero-order approximations α^0, β^0. The details can be found in Gayen et al. (2006).

This process can be repeated in principle to obtain higher order solutions. However, Gayen et al. (2006) did not pursue this further as the first order solutions produced sufficiently accurate approximations for some quantities of physical interest involved in the water wave problem studied.

9.3 SOLUTION OF THE COUPLED INTEGRAL EQUATIONS FOR ANY l

For the values of $a(x), f_1(x), f_2(x)$ given by (9.2.1), (9.2.2), (9.2.3) the coupled integral equations (9.1) and (9.2) have been solved by Gayen et al (2007) *for any l*. For this purpose, the equations (9.1) and (9.2) have been written in the operator form

$$(S\varphi_1)(x) + (N\varphi_2)(x) = f_1(x), \ x>0, \qquad (9.3.1)$$

$$(S\varphi_2)(x) + (N\varphi_1)(x) = f_2(x), \ x>0, \qquad (9.3.2)$$

where the singular operator S and the non-singular operator N are defined by

$$(S\varphi)(x) = a(x) \varphi(x) + \frac{1}{\pi} \int_0^\infty \frac{\varphi(t)}{t-x} dt, \ x>0, \qquad (9.3.3a)$$

$$(N\varphi)(x) = -\frac{1}{\pi} \int_0^\infty \frac{\varphi(t)e^{-lt}}{t+x} dt, \ x > 0. \qquad (9.3.3b)$$

It is observed that the Carleman singular integral equation

$$(S\varphi)(x) = f(x), \ x > 0 \qquad (9.3.4)$$

has the explicit solution

$$\varphi(x) = \left(S^{-1} f\right)(x)$$

$$= \frac{\Phi_0^+(x)}{a(x) - i} \left(\hat{S}h\right)(x), \ x > 0 \tag{9.3.5}$$

with

$$h(t) = \frac{f(t)}{\Phi_0^+(t)\left(a(t) - i\right)}, \ t > 0 \tag{9.3.6}$$

where the operator $\hat{S}$ is defined by

$$\left(\hat{S}h\right)(x) = a(x)\,h(x) - \frac{1}{\pi} \int_0^\infty \frac{h(t)}{t - x}\, dt, \ x > 0 \tag{9.3.7}$$

and

$$\Phi_0^+(x) = \lim_{y \to +0} \Phi_0(z), \ z = x + iy,$$

where

$$\Phi_0(z) = \exp\left[\frac{1}{2\pi i}\left\{\int_0^\infty \left(\ln\frac{t - iK_1}{t + iK_1} - 2\pi i\right)\frac{dt}{t - z} - \int_0^\infty \left(\ln\frac{t - iK_2}{t + iK_2} - 2\pi i\right)\frac{dt}{t - z}\right\}\right], \tag{9.3.8}$$

$$z \notin (0, \infty)$$

is a solution of the homogeneous problem

$$\left(a(x) + i\right)\Phi_0^+(x) - \left(a(x) - i\right)\Phi_0^-(x) = 0, \ x > 0. \tag{9.3.9}$$

We now apply the operator S^{-1} to (9.3.1) to obtain

$$\varphi_1(x) = S^{-1}\left(f_1 - \mathcal{N}\varphi_2\right)(x), \ x > 0 \tag{9.3.10}$$

which when substituted into (9.3.2) produces

$$\left(S\varphi_2\right)(x) + \mathcal{N}\left(S^{-1}\left(f_1 - \mathcal{N}\varphi_2\right)\right)(x) = f_2(x), \ x > 0. \tag{9.3.11}$$

Applying the operator S^{-1} to both sides of (9.3.11), we find

$$\left(\left(\mathcal{I}-\mathcal{L}^2\right)\varphi_2\right)(x)=r(x),\ x>0 \tag{9.3.12}$$

where $\mathcal{I}$ is the identity operator and

$$\mathcal{L}=S^{-1}\mathcal{N}, \tag{9.3.13}$$

$$r(x)=S^{-1}\left(f_2-\mathcal{N}S^{-1}f_1\right)(x),\ x>0. \tag{9.3.14}$$

It should be noted that the operator $\mathcal{N}S^{-1}$ is not commutative.

Now $f_1(x)$ and $f_2(x)$ are substituted from (9.2.2) and (9.2.3) into (9.3.14) to obtain $r(x)$ in the form

$$r(x)=\alpha\ r_1(x)+\beta\ r_2(x)$$

where

$$r_1(x)=\frac{1}{2c}\frac{\Phi_0^+(x)}{a(x)-1}\left[\frac{1}{x+iK_2}+\frac{e^{iK_2l}}{\pi}\int_0^\infty\frac{\Phi_0^+(t)e^{-lt}}{(a(t)-i)(t+x)(t-iK_2)\Phi_0(-t)}dt\right] \tag{9.3.15}$$

and

$$r_2(x)=\frac{1}{2c}\frac{\Phi_0^+(x)}{a(x)-1}\left[\frac{e^{iK_2l}}{x-iK_2}+\frac{1}{\pi}\int_0^\infty\frac{\Phi_0^+(t)e^{-lt}}{(a(t)-i)(t+x)(t+iK_2)\Phi_0(-t)}dt\right] \tag{9.3.16}$$

with

$$c=\Phi_0\left(\pm iK_2\right)=\left(\frac{2K_2}{K_1+K_2}\right)^{1/2}.$$

We now define two functions $\psi(x)$ and $\chi(x)$ for $x>0$ such that

$$\left(\left(\mathcal{I}+S\right)\varphi_2\right)(x)=\psi(x),\ \left(\left(\mathcal{I}-S\right)\varphi_2\right)(x)=\chi(x),\ x>0 \tag{9.3.17}$$

so that

$$\varphi_2(x)=\frac{1}{2}\left(\psi(x)+\chi(x)\right),\ \left(S\varphi_2\right)(x)=\frac{1}{2}\left(\psi(x)-\chi(x)\right),\ x>0 \tag{9.3.18}$$

Then the integral equation (9.3.12) can be written either as

$$((\mathcal{I}+\mathcal{L})\chi)(x) = r(x), \ x > 0 \qquad (9.3.19)$$

or as

$$((\mathcal{I}-\mathcal{L})\psi)(x) = r(x), \ x > 0. \qquad (9.3.20)$$

Since

$$r(x) = \alpha \ r_1(x) + \beta \ r_2(x)$$

we may express $\psi(x)$, $\chi(x)$ as

$$\psi(x) = ((\mathcal{I}-\mathcal{L})^{-1}r)(x) = \alpha \ \psi_1(x) + \beta \ \psi_2(x), \qquad (9.3.21)$$

$$\chi(x) = ((\mathcal{I}+\mathcal{L})^{-1}r)(x) = \alpha \ \chi_1(x) + \beta \ \chi_2(x), \qquad (9.3.22)$$

where $\psi_j(x)$, $\chi_j(x)$ $(j=1,2)$, $x > 0$ are unknown functions.

The integral equation (9.3.19) along with the relation (9.3.21), and the integral equation (9.3.20) along with the relation (9.3.22) are satisfied if $\psi_j(x)$, $\chi_j(x)$ $(j=1,2)$ satisfy

$$((\mathcal{I}-\mathcal{L})\psi_1)(x) = r_1(x), \ x > 0, \quad ((\mathcal{I}-\mathcal{L})\psi_2)(x) = r_2(x), \ x > 0,$$
$$((\mathcal{I}+\mathcal{L})\chi_1)(x) = r_1(x), \ x > 0, \quad ((\mathcal{I}+\mathcal{L})\chi_2)(x) = r_2(x), \ x > 0. \qquad (9.3.23)$$

These are in fact Fredholm integral equations with regular kernels.

The integral operator $\mathcal{L}$

The integral operator $\mathcal{L} = \mathcal{S}^{-1}\mathcal{N}$ is now obtained explicitly. Using the definitions of the integral operators $\mathcal{S}^{-1}$ and $\mathcal{N}$ as given in (9.3.5) and (9.3.3b) respectively, it is easy to see that

$$
\begin{aligned}
(\mathcal{L}m)(x) &= ((\mathcal{S}^{-1}\mathcal{N})m)(x) \\
&= \frac{\Phi_0^+(x)}{a(x)-i}\left[\frac{a(x)}{\Phi_0^+(x)(a(x)+i)}\left(-\frac{1}{\pi}\int_0^\infty \frac{m(t)e^{-lt}}{t+x}dt \right) \right. \\
&\quad \left. +\frac{1}{\pi^2}\int_0^\infty m(t)\, e^{-lt}\left(\int_0^\infty \frac{d\xi}{\Phi_0^+(\xi)(a(\xi)+i)(\xi+t)(\xi-x)} \right)dt \right].
\end{aligned}
\qquad (9.3.24)
$$

To evaluate the inner integral in the second term of (9.3.24), we consider the integral

$$\int_\Gamma \frac{d\zeta}{\Phi_0(\zeta)(\zeta - z + t)(\zeta - z)}, \quad z \notin \Gamma \qquad (9.3.25)$$

where Γ is a positively oriented contour consisting of a loop around the positive real axis of the complex ζ-plane, having indentations above the point $\zeta = x + i0$ and below the point $\zeta = x - i0$ in the complex ζ-plane. Also $\Phi_0(\zeta)$ satisfies the homogeneous RHP

$$\big(a(\xi) + i\big)\Phi_0^+(\xi) - \big(a(\xi) - i\big)\Phi_0^-(\xi) = 0, \quad \xi > 0. \qquad (9.3.26)$$

We observe that

$$\int_\Gamma \frac{d\zeta}{\Phi_0(\zeta)(\zeta + t)(\zeta - z)} = \int_0^\infty \left\{ \frac{1}{\Phi_0^+(\xi)} - \frac{1}{\Phi_0^-(\xi)} \right\} \frac{d\xi}{(\xi + t)(\xi - z)} \qquad (9.3.27)$$

$$= 2i \int_0^\infty \frac{d\xi}{\Phi_0^+(\xi)\big(a(\xi) + i\big)(\xi + t)(\xi - z)}.$$

Also from the residue calculus theorem

$$\int_\Gamma \frac{d\zeta}{\Phi_0(\zeta)(\zeta + t)(\zeta - z)} = \frac{2\pi i}{t + z}\left\{ \frac{1}{\Phi_0(z)} - \frac{1}{\Phi_0(-t)} \right\}. \qquad (9.3.28)$$

Comparing (9.2.27) and (9.2.28) we find

$$\frac{1}{t + z}\left\{ \frac{1}{\Phi_0(z)} - \frac{1}{\Phi_0(-t)} \right\} = \frac{1}{2\pi i} \int_0^\infty \frac{2i \, d\xi}{\Phi_0^+(\xi)\big(a(\xi) + i\big)(\xi + t)(\xi - z)}.$$

Applying Plemelj formulae to the above relation, the inner integral in the second term on the right side of (9.3.24) is evaluated as

$$\int_0^\infty \frac{d\xi}{\Phi_0^+(\xi)\big(a(\xi) + i\big)(\xi + t)(\xi - x)} = \frac{\pi}{t + x}\left\{ \frac{a(x)}{\big(a(x) + i\big)\Phi_0^+(x)} - \frac{1}{\Phi_0(-t)} \right\}$$

which when substituted into (9.3.24), produces

$$(\mathcal{L}m)(x) = -\frac{1}{\pi}\frac{\Phi_0^+(x)}{a(x) - i}\int_0^\infty \frac{m(t)e^{-lt}}{(t + x)\Phi_0(-t)}dt. \qquad (9.3.29)$$

The Fredholm integral equations in (9.3.23) can be solved numerically and then the functions $\psi_j(x)$, $\chi_j(x)$ ($j = 1, 2$) can be found numerically. It may be noted that considerable analytical calculations are required to reduce the functions $r_j(x)$ ($j = 1, 2$) to forms suitable for numerical computation.

Simplification of $r_1(x)$ and $r_2(x)$

The basic step for the evaluation of the integral equations (9.3.23) and the functions $r_1(x)$, $r_2(x)$ is to determine the functions $\Phi_0(-x)$ and $\Phi_0^+(x)$ for $x > 0$ in computable forms. Now an explicit derivation of these functions and simplification of $r_1(x)$ and $r_2(x)$ is given.

The function $\Phi_0^+(x)$ is given by

$$\Phi_0^+(x) = \left(\frac{x - iK_1}{x + iK_1} \frac{x + iK_2}{x - iK_2} \right)^{1/2} \exp \left(\frac{1}{2\pi i} \int_0^\infty \frac{\ln\left(\frac{t - iK_1}{t + iK_1} \frac{t + iK_2}{t - iK_2} \right)}{t - x} dt \right), \quad x > 0. \quad (9.3.30)$$

If we define

$$Y(x) = \frac{1}{2\pi i} \int_0^\infty \frac{\ln\left(\frac{t - iK_1}{t + iK_1} \frac{t + iK_2}{t - iK_2} \right)}{t - x} dt, \quad x > 0,$$

$$Y_j(x) = \frac{1}{2\pi i} \int_0^\infty \ln \frac{t - iK_j}{t + iK_j} \frac{dt}{t - x} \quad (j = 1, 2), \quad x > 0, \quad (9.3.31)$$

$$X(x) = Y(-x) \quad \text{and} \quad X_j(x) = Y_j(-x), \quad x > 0,$$

then

$$Y(x) = Y_1(x) - Y_2(x), \quad \Phi_0(-x) = \exp(X(x)),$$

$$\Phi_0^+(x) = \left(\frac{x - iK_1}{x + iK_1} \frac{x + iK_2}{x + iK_2} \right)^{1/2} \exp(Y(x)). \quad (9.3.32)$$

Following Varley and Walker (1989) the derivative of $Y'_j(x)$ is found to be

$$Y'_j(x) = -\frac{K_j}{2\pi}\left[\frac{\ln\left(x/(-iK_j)\right)}{x\left(x+iK_j\right)}+\frac{\ln\left(x/(iK_j)\right)}{x\left(x-iK_j\right)}\right], \quad j=1,2.$$

It may be observed that $Y_j(\infty) = 0$. Integration of $Y'_j(x)$ gives

$$Y_j(x) = -\frac{K_j}{2\pi}\left[\int_{\infty}^{x}\frac{\ln\left(t/(-iK_j)\right)}{t\left(t+iK_j\right)}+\frac{\ln\left(t/(iK_j)\right)}{t\left(t-iK_j\right)}\right]dt \tag{9.3.33}$$

$$= -\frac{1}{2\pi i}\int_{-iK_j/x}^{iK_j/x}\frac{\ln u}{u-1}\,du.$$

After some manipulations $Y(x)$ reduces to

$$Y(x) = \frac{1}{4}\frac{x-iK_1}{x-iK_2}-\frac{3}{4}\frac{x+iK_1}{x+iK_2}-\frac{1}{\pi}\int_{K_2/x}^{K_1/x}\frac{\ln v}{v^2+1}\,dv. \tag{9.3.34}$$

Hence $\Phi_o^+(x)$ has the alternative form

$$\Phi_o^+(x) = \left(\frac{x-iK_1}{x-iK_2}\right)^{1/2}\left(\frac{x+iK_1}{x+iK_2}\right)^{-1/2}\exp\left(Y(x)\right)$$

$$= \left(\frac{x-iK_1}{x-iK_2}\right)^{3/4}\left(\frac{x+iK_1}{x+iK_2}\right)^{-5/4}E(x) \tag{9.3.35}$$

$$= \left(\frac{x^2+K_1^2}{x^2+K_2^2}\right)^{-1/4}e^{-2i(\theta_1-\theta_2)}E(x)$$

where

$$\theta_j = \tan^{-1}\left(K_j/x\right), j=1,2 \text{ and} \tag{9.3.36}$$

$$E(x) = \exp\left(-\frac{1}{\pi}\int_{K_2/x}^{K_1/x}\frac{\ln v}{v^2+1}\,dv\right).$$

$X(x)$ is simplified in a similar manner and we find that

$$X_j(x) = Y_j(-x) = -Y_j(x), j=1,2.$$

Thus $X(x) = -Y(x)$, and

$$\Phi_0(-x) = \exp(X(x))$$

$$= \left(\frac{x - iK_1}{x - iK_2}\right)^{-1/4} \left(\frac{x + iK_1}{x + iK_2}\right)^{3/4} (E(x))^{-1} \qquad (9.3.37)$$

$$= \left(\frac{x^2 + K_1^2}{x^2 + K_2^2}\right)^{1/4} e^{i(\theta_1 - \theta_2)} (E(x))^{-1}.$$

The various complex-valued functions appearing in $r_1(x)$ and $r_2(x)$ are simplified as follows:

(a) $\dfrac{\Phi_0^+(x)}{a(x) - i} = (K_1 - K_2)x(x^2 + K_1^2)^{-3/4}(x^2 + K_2^2)^{-1/4} e^{-i(\theta_1 - \theta_2)} E(x)$

where we have used

$$a(x) - i = \frac{(x - iK_1)(x + iK_2)}{x(K_1 - K_2)}.$$

(b) $\dfrac{\Phi_0^+(x)}{a(x) - i} \dfrac{1}{\Phi_0(-x)} = \dfrac{(K_1 - K_2)x}{x^2 + K_1^2} e^{-2i(\theta_1 - \theta_2)} (E(x))^2.$

(c)

$$\dfrac{\Phi_0^+(x)}{a(x) - i} \dfrac{1}{\Phi_0(-x)} \left(\dfrac{1}{x - iK_2}, \dfrac{1}{x + iK_2}\right)$$

$$= \dfrac{(K_1 - K_2)x \, e^{-2i\theta_1}}{(x^2 + K_1^2)(x^2 + K_2^2)^{1/2}} = (E(x))^2 \left(e^{3i\theta_2}, e^{i\theta_2}\right).$$

(d) $\dfrac{\Phi_0^+(x)}{a(x) - i}\left(\dfrac{1}{x - iK_2}, \dfrac{1}{x + iK_2}\right) = \dfrac{(K_1 - K_2)x \, e^{-i\theta_1}}{\{(x^2 + K_1^2)(x^2 + K_2^2)\}^{3/4}} E(x)\left(e^{2i\theta_2}, 1\right).$

Using (a) to (d), $r_1(x)$ and $r_2(x)$ are simplified as

$$r_1(x) = r_0(x)\left[\frac{1}{\left(x^2 + K_2^2\right)^{1/2}} + e^{iK_2l}\int_0^\infty M(u,x)\, e^{3i\,\tan^{-1}(K_2/u)}du\right],$$

$$r_2(x) = r_0(x)\left[\frac{e^{i(K_2l+2\theta_2)}}{\left(x^2 + K_2^2\right)^{1/2}} + \int_0^\infty M(u,x)\, e^{i\,\tan^{-1}(K_2/u)}du\right],$$

where

$$r_0(x) = \frac{K_1 - K_2}{\pi}\, \frac{x\, E(x)\, e^{-i\theta_1}}{\left(x^2 + K_1^2\right)^{3/4}}\left(x^2 + K_2^2\right)^{1/4},$$

$$\tag{9.3.38}$$

$$M(u,x) = \frac{K_1 - K_2}{2c}\, \frac{u\, e^{-lu}\,\left(E(x)\right)^2\, e^{i\theta_2 - 2i\,\tan^{-1}(K_1/u)}}{\left(u^2 + K_1^2\right)\left(u^2 + K_2^2\right)^{1/2}(u+x)}.$$

The functions $\varphi_1(x), \varphi_2(x)$

The functions $\varphi_1(x)$ and $\varphi_2(x)$ which satisfy the two coupled singular integral equations (9.3.1) and (9.3.2) are now found in a straight forward manner as

$$\varphi_1(x) = \left(S^{-1}f_1\right)(x) - \left(\mathcal{L}\varphi_2(x)\right)$$

$$= \left(S^{-1}f_1\right)(x) - \frac{1}{2}\{\psi(x) - \chi(x)\} \tag{9.3.39}$$

$$= \alpha\, \varphi_1^\alpha(x) + \beta\, \varphi_1^\beta(x),$$

$$\varphi_2(x) = \frac{1}{2}\{\psi(x) + \chi(x)\}$$

$$= \alpha\, \varphi_2^\alpha(x) + \beta\, \varphi_2^\beta(x) \tag{9.3.40}$$

where

$$\varphi_1^\alpha(x) = \frac{1}{2}\left[\frac{\Phi_0^+(x)e^{iK_2l}}{c(a(x)-i)(x-iK_2)} - \psi_1(x) + \chi_1(x)\right], \tag{9.3.41}$$

$$\varphi_1^\beta(x) = \frac{1}{2}\left[\frac{\Phi_0^+(x)}{c(a(x)-i)(x+iK_2)} - \psi_2(x) + \chi_2(x)\right], \quad (9.3.42)$$

$$\varphi_2^\alpha(x) = \frac{1}{2}(\psi_1(x) + \chi_1(x)) \quad\quad (9.3.43)$$

$$\varphi_2^\beta(x) = \frac{1}{2}(\psi_2(x) + \chi_2(x)); \quad\quad (9.3.44)$$

the value of the constant c being given earlier.

Thus $\varphi_1(x)$ and $\varphi_2(x)$ are obtained in principle for *any value* of the parameter l.

Remarks

1. The integral equations (9.3.3a) and (9.3.3b) are coupled. They can be decoupled simply by addition and subtraction in the following manner. If we define

$$\varphi(x) = \varphi_1(x) + \varphi_2(x), \quad \psi(x) = \psi_1(x) - \psi_2(x) \quad\quad (9.3.45)$$

then addition and subtraction of equations (9.3.3a) and (9.3.3b) produce

$$a(x)\,\varphi(x) + \frac{1}{\pi}\int_0^\infty \frac{\varphi(t)}{t-x}\,dt - \frac{1}{\pi}\int_0^\infty \frac{\varphi(t)e^{-lt}}{t+x}\,dt = f(x), \quad x>0 \quad (9.3.46)$$

$$a(x)\,\psi(x) + \frac{1}{\pi}\int_0^\infty \frac{\psi(t)}{t-x}\,dt + \frac{1}{\pi}\int_0^\infty \frac{\psi(t)e^{-lt}}{t+x}\,dt = g(x), \quad x>0 \quad (9.3.47)$$

where

$$f(x), g(x) = f_1(x) \pm f_2(x). \quad\quad (9.3.48)$$

The two equations (9.3.46) and (9.3.47) are not coupled. A similar approach as described above can be employed to solve them for *any value* of l. This is described below briefly.

Using the operators S and N defined in (9.3.3a) and (9.3.3b) respectively, the equations (9.3.46) and (9.3.47) reduce to

$$(S\varphi)(x)+(N\psi)(x)=f(x),\quad x>0,\qquad\qquad(9.3.49)$$

$$(S\varphi)(x)+(N\psi)(x)=g(x),\quad x>0.\qquad\qquad(9.3.50)$$

Applying the operator S^{-1} to the above equations we find that

$$((I+L)\varphi)(x)=(S^{-1}f)(x),\quad x>0,\qquad\qquad(9.3.51)$$

$$((I+L)\psi)(x)=(S^{-1}g)(x),\quad x>0,\qquad\qquad(9.3.52)$$

where the operator L is defined in (9.3.29).

The right-hand sides of (9.3.51) and (9.3.52) are of the forms

$$(S^{-1}f)(x)=\alpha\,f^{\alpha}(x)+\beta\,f^{\beta}(x)\qquad\qquad(9.3.53)$$

$$(S^{-1}g)(x)=\alpha\,g^{\alpha}(x)+\beta\,g^{\beta}(x)\qquad\qquad(9.3.54)$$

so that

$$\begin{aligned}\varphi(x)&=(I+L)^{-1}(\alpha\,f^{\alpha}+\beta\,f^{\beta})(x)\\&=\alpha\,\varphi^{\alpha}(x)+\beta\,\varphi^{\beta}(x)\end{aligned}\qquad(9.3.55)$$

and

$$\begin{aligned}\psi(x)&=(I-L)^{-1}(\alpha\,g^{\alpha}+\beta\,g^{\beta})(x)\\&=\alpha\,\psi^{\alpha}(x)+\beta\,\psi^{\beta}(x),\end{aligned}\qquad(9.3.56)$$

where $\varphi^{\alpha}(x),\varphi^{\beta}(x),\ \psi^{\alpha}(x),\psi^{\beta}(x)$ are to be found.

Comparing (9.3.51) and (9.3.52) (together with (9.5.53) and (9.5.54)) to (9.3.55) and (9.3.56) we see that equations (9.3.51) and (9.3.52) will be satisfied if the functions $\varphi^{\alpha}(x),\varphi^{\beta}(x),\ \psi^{\alpha}(x),\psi^{\beta}(x)$ satisfy the following Fredholm integral equations of the second kind.

$$((I+L)\varphi^{\alpha})(x)=f^{\alpha}(x),\ x>0,\ ((I+L)\varphi^{\beta})(x)=f^{\beta}(x),\ x>0$$
$$((I-L)\varphi^{\alpha})(x)=g^{\alpha}(x),\ x>0,\ ((I-L)\psi^{\beta})(x)=g^{\beta}(x),\ x>0.\qquad(9.3.57)$$

Once the four equations (9.3.57) are solved numerically the functions $\varphi(x), \psi(x)$ can be found in terms of α and β using (9.3.55) and (9.3.56), and then $\varphi_1(x)$ and $\varphi_2(x)$ can be determined from (9.3.45).

2. The form $a(x) = \dfrac{x^2(Dx^2 + 1) + K^2}{DKx^2}$, and some specific forms of $f_1(x)$ and $f_2(x)$

   (different from (9.2.2) and (9.2.3)) occur in a water wave problem involving a floating elastic plate of finite width l in two dimensions (see Gayen and Mandal 2009). The coupled, singular integral equations in this situation can be solved by reducing them to twelve Fredholm integral equations of second kind with regular kernels. Details can be found in Gayen and Mandal (2009).

3. The forms of the coupled singular integral equations of Carleman type given by (9.1) and (9.2) can be generalized to the forms

$$(S_1\varphi_1)(x) + (N\varphi_2)(x) = f_1(x), \quad x > 0, \qquad (9.3.58)$$

$$(S_2\varphi_2)(x) + (N\varphi_2)(x) = f_2(x), \quad x > 0, \qquad (9.3.59)$$

where

$$(S_i\varphi)(x) = a_i(x)\varphi(x) + \frac{1}{\pi}\int_0^\infty \frac{\varphi(t)}{t - x}\,dt, x > 0,\ i = 1, 2$$

and N is the same as defined in (9.3.3b).

Finding the method of solution of this pair of equations for *any value* of the parameter l appears to be a challenging task.

Bibliography

1. Atkinson, K. E., *Introduction to Numerical Analysis,* John Wiley, New York, 1989.
2. Atkinson, K. E., *The Numerical Solution of Integral Equations of the Second Kind,* Cambridge University Press, 1997.
3. Baker, C. T. H., *The Numerical Treatment of Integral Equations,* Clarendon Press, Oxford, 1978.
4. Banerjea, S., Dolai, D. P. and Mandal, B. N., On waves due to rolling of a ship in water of finite depth. *Arch. Appl. Mech.* **67** (1996) 35–45.
5. Banerjea, S. and Mandal, B. N., Solution of a singular integral equation in double interval arising in the theory of water waves. *Appl. Math. Lett.* **6** no.3 (1993) 81–84.
6. Banerjea, S. and Rakshit, P., Solution of a singular integral equation with log kernel. *Int. J. Appl. Math. & Engg. Sci.* **1** (2007) 297–301.
7. Bhattacharya, S. and Mandal, B. N., Numerical solution of a singular integro-differential equation. *Appl. Math. Comp.* **195** (2008) 346–350.
8. Bhattacharya, S. and Mandal, B. N., Numerical solution of some classes of logarithmically singular integral equations. *J. Adv. Res. Appl. Math.* **2** (2010) 30–38.
9. Bhattacharya, S. and Mandal, B. N., Numerical solution of an integral equation arising in the problem of cruciform crack. *Int. J. Appl. Math. & Mech.* **6** (2010) 70–77.
10. Boersma, J., Note on an integral equation of viscous flow theory. *J. Engg. Math.* **12** (1978) 237–243.
11. Brown, S. N., On an integral equation of viscous flow theory. *J. Engg. Math.* **11** (1977) 219–226.
12. Capobianco, M. R. and Mastronardi, N., A numerical method for Volterra-type integral equation with logarithmic kernel. *Facta Universitates (NIS) Ser. Math. Inform.* **13** (1998) 127–138.
13. Chakrabarti, A., Derivation of the solution of certain singular integral equations. *J. Indian Inst. Sci.* **62B** (1980) 147–157.
14. Chakrabarti, A., Solution of two singular integral equations arising in water wave problems. *Z. Angew. Math. Mech.* **69** (1989) 457–459.
15. Chakrabarti, A., A note on singular integral equations. *Int. J Math. Educ. Sci. Technol.* **26** (1995) 737–742.
16. Chakrabarti, A., A survey of two mathematical methods used in scattering of surface waves, in Mathematical Techniques for Water Waves, *Advances in Fluid Mech.* (ed. Mandal, B. N.), Computational Mechanics Publications, Southampton, **8** (1997) 231–253.
17. Chakrabarti, A., On the solution of the problem of scattering of surface waer waves by a sharp discontinuity in the surface boundary conditions. *ANZIAM J.* **42** (2000) 277–286.

18. Chakrabarti, A., Solution of certain weakly singular integral equations. *IMA J. Appl. Math.* **71** (2006) 534–543.

19. Chakrabarti, A., Solution of a simple hypersingular integral equation. *J. Integr. Eqns. & Applic.* **19** (2007) 465–472.

20. Chakrabarti, A., Solution of a generalized Abel integral equation. *J. Integr. Eqns. & Applic.* **20** (2008) 1–11.

21. Chakrabarti, A., *Applied Integral Equations.* Vijay Nicole Imprints Pvt. Ltd. Chennai, 2008.

22. Chakrabarti, A. and George, A. J., Solution of a singular integral equation involving two intervals arising in the theory of water waves. *Appl. Math. Lett.* **7** no. 5 (1994) 43–47.

23. Chakrabarti, A. and George, A. J., Diagonizable Abel integral operators. *SIAM J. Appl. Math.* **57** (1997) 568–575.

24. Chakrabarti, A. and Hamsapriye, Numerical solution of a singular integro-differential equation. *Z. Angew. Math. Mech.* **79** (1999) 233–241.

25. Chakrabarti, A., Manam, S. R. and Banerjea, S., Scattering of surface water waves in volving a vertical barrier with a gap. *J. Engg. Math.* **45** (2003) 183–194.

26. Chakrabarti, A. and Mandal, B. N., Derivation of the solution of a simple hypersingular integral equation. *Int. J. Math. Educ. Sci. Technol.* **29** (1998) 47–53.

27. Chakrabarti, A., Mandal, B. N., Basu, U. and Banerjea, S., Solution of a class of hypersingular integral equations of the second kind. *Z. Angew. Math. Mech.* **77** (1997) 319–320.

28. Chakrabarti, A. and Martha, S., Approximate solution of Fredholm integral equations of the second kind. *Appl. Math. Comp.* **211** (2009) 459–466.

29. Chakrabarti, A. and Sahoo, T., A note on a class of singular integro-differen-tial equations. *Proc. Indian Acad. Sci.* (Math. Sci.) **106** (1996) 177–182.

30. Chakrabarti, A. and Williams, W. E., A note on a singular mintegral equation. *J. Inst. Math. Applic.* **26** (1980) 321–323.

31. Chakrabarti, A. and Vanden Berghe, G., Approximate solution of singular integral equations. *Appl. Math. Lett.* **17** (2004) 553–559.

32. Chan, Y. S., Fannijiang, A. C. and Paulino, G. H., Integral equations with hypersingular kernels—theory and applications to fracture mechanics. *Int. J. Engg. Sci.* **41** (2003) 683–720.

33. Das, P., Banerjea, S. and Mandal, B. N., Scattering of oblique wavesby a thin vertical wall with a submerged gap. *Arch. Mech.* **48** (1996) 959–972.

34. Das, P., Dolai, D. and Mandal, B. N., Oblique water wave diffraction by two parallel thin barriers with gaps. *J. Wtry. Port Coast Ocean Engg.* **123** (1997) 163–171.

35. De, S., Mandal, B. N. and Chakrabarti, A., Water wave scattering by two submerged plane vertical barriers—Abel integral equations approach. *J. Engg. Math.* **65** (2009) 75–87.

36. De, S., Mandal, B. N. and Chakrabarti, A., Use of Abel integral equations in water wave scattering by two surface-piercing barriers. *Wave Motion* **47** (2010) 279–288.

37. Doetsch, G., *Introduction to the Theory and Application of the Laplace Trans-formation.* Springer-Verlag, Berlin, 1974.

38. Dragos, L. A collocation method for the integration of Prandtl's equation. *Z. Angew. Math. Mech.***74** (1994) 289–290.

39. Dragos, L., Integration of Prandtl's equation by the aid of quadrature formulae of Gauss. *Quart. Appl. Math.* **52** (1994) 23–29.

40. Elliot, D., The cruciform crack problem and sigmoidal transformations. *Math. Meth. Appl. Sci.* **20** (1997) 121–132.
41. Evans, D. V. and Morris, C. A. N., The effect of a fixed vertical barrier on obliquely incident surface waves in deep water. *J. Inst. Math. Applic.* **9** (1972) 198–204.
42. Estrada, R. and Kanwal, R. P., The Cauchy type integral equations. *Ganita* **36** (1985a) 1–23.
43. Estrada, R. and Kanwal, R. P., Distributional solutions of singular integral equations. *J. Integr. Eqns.* **8** (1985b) 41–85.
44. Estrada, R. and Kanwal, R. P., The Caleman singular integral equations. *SIAM Rev.* **29** (1987) 263–290.
45. Estrada, R. and Kanwal, R. P., Integral equations with logarithmic kernels. *IMA J. Appl. Math.* **42** (1989) 133–155.
46. Estrada, R. and Kanwal, R. P., *Singular Integral Equations*, Birkhauser, Boston, 2000.
47. Frankel, J. I., A Galerkin solution to a regularized Cauchy singular integro- differential equation. *Quart. J. Appl. Math.* **53** (1995) 245–258.
48. Fox, C., A generalization of the Cauchy principal value. *Canad. J. Math.* **9** (1957) 110–117.
49. Gakhov, F. D., *Boundary Value Problems*, Pergamon Press, Oxford, 1966.
50. Gayen, R. and Mandal, B. N., Scattering of surface water waves by a floating elastic plate in two dimensions. *SIAM J. Appl. Math.* **69** (2009) 1520–1541.
51. Gayen, R. and Mandal, B. N. and Chakrabarti, A., Water wave scattering by two sharp discontinuities in the surface boundary conditions. *IMA J. Appl. Math.* **71** (2006) 811–831.
52. Gayen, R. and Mandal, B. N. and Chakrabarti, A., Water wave diffraction by a surface strip. *J. Fluid Mech.* **571** (2007) 419–438.
53. Golberg, M. A., The convergence of several algorithms for solving integral equations with finite-part integrals: II. *J. Integral Eqns.* **9** (1985) 267–275.
54. Golberg, M. A. and Chen, C. S., *Discrete Projection Method for Integral Equations.* Comp. Mech. Publ., Southampton, 1997.
55. Hadamard, J. *Lectures on Cauchy's Problem in Linear Partial Differential Equations.* New York, 1952.
56. Holford, R. L., Short surface waves in the *presence* of a finite dock: I, II. *Proc. Camb. Phil. Soc.* **60** (1964) 957–983, 985–1011.
57. Jones, D. S., *The Theory of Electromagnetism.* Pergamon Press, London, 1964.
58. Jones, D. S., *Generalized Functions.* McGraw Hill, London, 1966.
59. Kanoria, M. and Mandal, B. N., Water wave scattering by a submerged circular-arc-shaped plate. *Fluid Dyn. Res.* **31** (2002) 317–331.
60. Khater, A. H., Shamardan, A. B., Callebaut, D. K. and Sakran, M. R. A., Legendre solutions of integral equations with logarithmic kernels. *Int. J. Computer Math.* **85** (2008) 53–63.
61. Knill, O., Diagonalization of Abel's integral operator. *SIAM J. Appl. Math.* **54** (1994) 1250–1253.
62. Kyth, P. K. and Puri, P., *Computational Methods for Linear Integral Equations.* Birkhauser, Boston, 2002.
63. Lewin, L., On the solution of a class of wave-guide discontinuity problems by the use of singular integral equations. *IRE Trans. MTT* **9** (1961) 321–332.

64. Lewin, L., *Theory of Wave-guides, Techniques for the Soltion of Waveguide Problems.* Newnes-Butterworths, London, 1975.
65. Lorenz, G. G., *Bernstein Polynomials.* Univ. of Toronto Press, Toronto, 1953.
66. M. Lowengrub. System of Abel type integral equations, Research Notes in Math., Edited by R. P. Gilbert and R. J. Weinacht, Pitman Publishing, 1976, 277–296.
67. Maleknejad, K., Lotfi, T. and Mahdiani, K., Numerical solution of first kind Fredholm integral equation with wavelets Galerkin method and wavelets precondition. *Appl. Math. Comp.* **186** (20070 794–800.
68. Maleknejad, K. and Mirazee, F., Numerical solution of linear Fredholm integral systemby rationalized Haar functions method. *Int. J. Computer. Math.* **80** (2003) 1397–1405.
69. Maleknejad, K. and Mirazee, F., Using rationalized Haar wavelet for solving linear integral equation. *Appl. Math. Comp.* **160** (2003) 579–587.
70. Maiti, P. and Mandal, B. N., Wave scattering by a thin vertical barrier submerged beneath an ice-cover in deep water. *Appl. Ocean Res.* **32** (2010) 367–373.
71. Mandal, B. N., Banerjea, S. and Dolai, P., Interface wave diffraction by a thin vertical barrier submerged in lower fluid. *Appl. Ocean Res.* **17** (1995) 93–102.
72. Mandal, B. N. and Bera, G. H., Approximate solution of a class of hypersingular integral equations. *Appl. Math. Lett.* **19** (2006) 1286–1290.
73. Mandal, B. N. and Bera, G. H., Approximate solution of a class of singular integral equations of second kind. *J. Comput. Appl. Math.* **206** (2007) 189–195.
74. Mandal, B. N. and Bhattacharya, S., Numerical slution of some classes of integral equations using Bernstein polynomials. *Appl. Math. Comp.* **190** (2007) 189–195.
75. Mandal, B. N. and Chakrabarti, A., On Galerkin method applicable to problems of water wave scattering by barriers. *Proc. Indian Natn. Sci. Acad.* **65A** (1999) 61–77.
76. Mandal, B. N. and Chakrabarti, A., *Water Wave Scattering by Barriers,* WIT Press, Southampton, 2000.
77. Mandal, B. N. and Das, P., Oblique diffractionof surface waves by a submerged vertical plate. *J. Engg. Math.* **30** (1996) 459–470.
78. Mandal, B. N. and Goswami, S. K., A note on a singular integral equation. *Indian J. pure appl. Math.* **14** (1983) 1352–1356.
79. Mandal, B. N. and Gayen (Chowdhury), R., Water wave scattering by two symmetric circular arc shaped plates. *J. Engg. Math.* **44** (2002) 297–303.
80. Mandal, N., Chakrabarti, A. and Mandal, B. N., Solution of a system of generalized Abel integral equations. *Appl. Math. Lett.* **9** (no.5)1–4.
81. Martin, P. A., End-point behaviour of solutions of hypersingular integral equations, *Proc. R. Soc. Lond.* **A 432** (1991) 301–320.
82. Martin, P. A., Exact solution of a simple hypersingular integral equation. *J. Integr. Eqns. & Applic.* **4** (1992) 197–204.
83. Martin, P. A., Parsons, N. F. and Farina, L., Interaction of water waves with thin plates, in Mathematical Techniques for Water Waves, *Advances in Fluid Mech.* (ed. Mandal, B. N.), Computational Mechanics Publications, Southampton, **8** (1997) 197–227.
84. Muskhelishvili, N. I., *Singular Integral Equations,* P. Noordhoff, Groningen, Netherlands, 1953.
85. Pandey, R. K. and Mandal, B. N., Numerical solution of a system of generalized Abel integral equations using Bernstein polynomials. *J. Adv. Res in Sci. Comput.* **2** (2010) 44–53.

86. Parsons, N. F. and Martin, P. A., Scattering of water waves by submerged plates using hypersingular integral equations. *Appl. Ocean Res.* **14** (1992) 313–321.

87. Parsons, N. F. and Martin, P. A., Scattering of water waves by submerged curved plates and surface-piercing flat plates. *Appl. Ocean Res.* **16** (1994) 129–139.

88. Peters, A. S., A note on the integral equation of the first kind with a Cauchy kernel. *Comm. Pure Appl. Math.* **16** (1963) 57–61.

89. Porter, D., The transmission of surface waves through a gap in a vertical barrier. *Proc. Camb. Phil. Soc.* **71** (1972) 411–421.

90. Reihani, M. H. and Abadi, Z., Rationalized Haar functions method for solving Fredholm integral equations. *J. Comput. Appl. Math.* **200** (2007) 12–20.

91. Rooke, D. P. and Sneddon, I. N., The crack energy and stress intensity factor for a cruciform crack deformed by internal pressure. *Int. J. Engg. Sci.* **7** (1969) 1079–1089.

92. Sloan, L. H., and Stephan, E. P., Collocation with Chebyshev polynomials for Symm's integral equation on an interval. *J. Austral. Math. Soc. Ser.* **B 34** (1992) 199–211.

93. Sneddon, I. N., *The Use of Integral Transforms*. McGraw Hill, New York, 1972.

94. Snyder, M. A., *Chebyshev Methods in Numerical Approximation*. Prentice Hall, Englewood Cliffs., New Jersy, 1969.

95. Spence, D. A., A flow past a thin wing with an oscillating jet flap. *Phil. Trans. R. Soc.* **257** (1960) 445–467.

96. Stallybrass, M. P., A pressurized crack in the form of a cross. *Quart. J. Appl. Math.* **23** (1970) 35–48.

97. Stewartson, K., A note on lifting line theory. *Q. J. Mech. Appl. Math.* **13** (1960) 49–56.

98. Tang, B. Q. and Li, X. F., Approximate solution to an integral equation with fixed singularity for a cruciform crack. *Appl. Math. Lett.* **21** (2007) 1238–1244.

99. Ursell, F., The effect of a fixed vertical barrier on surface waves in deep water. *Proc. Camb. Phil. Soc.* **43** (1947) 374–382.

100. Vainikko, G. and Uba, P., A piecewise polynomial approximation to the solution of an integral equation with weakly singular kernel. *J. Austral. Math. Soc. Ser.* **B 22** (1981) 431–438.

101. Varley, E. and walker, J. D. A., A method for solving singular integro-differential equations. *IMA J. Appl. Math.* **43** (1989) 11–45.

102. Walton, J. System of generalized Abel integral equations with applications to simultaneous dual relations. *SIAM J. Math. Anal.* **109** (1979) 808–822.

103. Williams, W. E., A note on a singular integral equation. *J. Inst. Math. Applic.* **22** (1978) 211–214.

104. Wolfersdorf, L. V., Monotonicity methods for nonlinear singular integral and integro-differential equations. *Z. Angew. Math. Mech.* **63** (1983) 249–259.

Subject Index

T - #0122 - 101024 - C0 - 234/156/15 [17] - CB - 9781578087105 - Gloss Lamination